I0112693

The Wake of HMS *Challenger*

The Wake of HMS *Challenger*

How a Legendary Victorian Voyage Tells the Story of Our Oceans' Decline

Gillen D'Arcy Wood

PRINCETON UNIVERSITY PRESS
PRINCETON & OXFORD

Copyright © 2025 by Princeton University Press

Princeton University Press is committed to the protection of copyright and the intellectual property our authors entrust to us. Copyright promotes the progress and integrity of knowledge created by humans. Thank you for supporting free speech and the global exchange of ideas by purchasing an authorized edition of this book. If you wish to reproduce or distribute any part of it in any form, please obtain permission.

Requests for permission to reproduce material from this work should be sent to permissions@press.princeton.edu

Published by Princeton University Press
41 William Street, Princeton, New Jersey 08540
99 Banbury Road, Oxford OX2 6JX

press.princeton.edu

All Rights Reserved

ISBN 978-0-691-23324-6
ISBN (e-book) 978-0-691-23325-3

Library of Congress Control Number: 2025936729

British Library Cataloging-in-Publication Data is available

Editorial: Ingrid Gnerlich and Whitney Rauenhorst
Production Editorial: Kathleen Cioffi
Text Design: Chris Ferrante
Production: Jacqueline Poirier
Publicity: Maria Whelan and Kate Farquhar-Thomson
Copyeditor: Susan Matheson

Jacket image: Alice Strange, *Flight Paths*, 2013, digital collage.

This book has been composed in Miller and Sweet Sans

Printed in the United States of America

1 3 5 7 9 10 8 6 4 2

No one could write truthfully about
the sea and leave out the poetry.

—RACHEL CARSON

CONTENTS

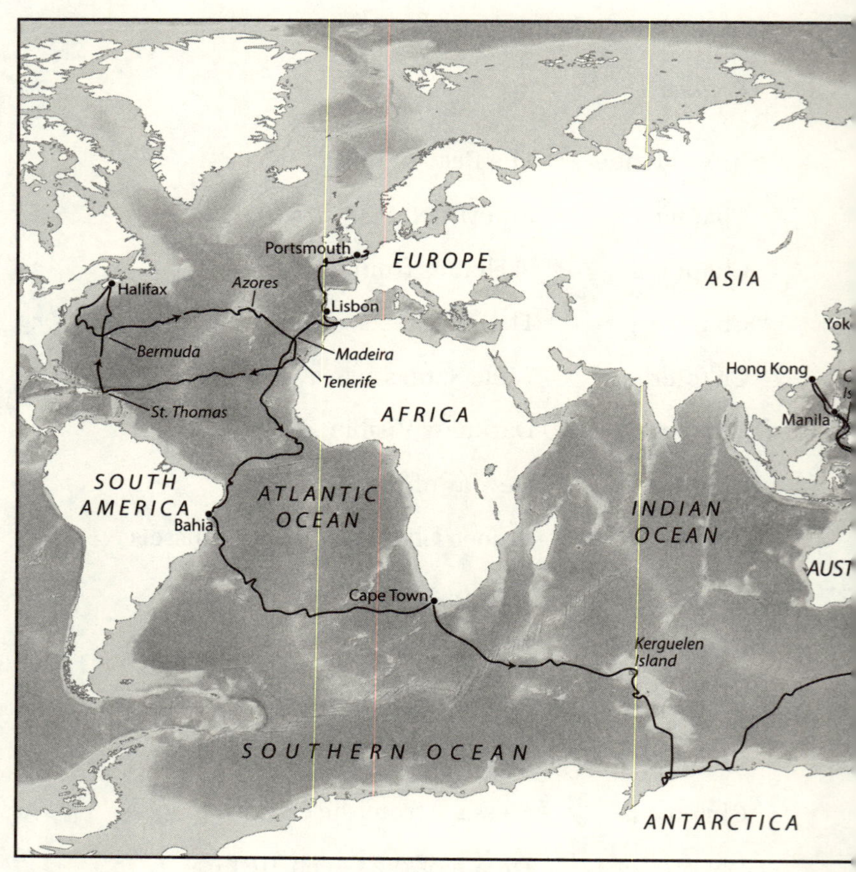

FIGURE 0.1 Map of HMS *Challenger* voyage. December 1872–May 1876. Adapted from William J. Spry, *The Cruise of Her Majesty's Ship Challenger* (London: Sampson Low, 1876).

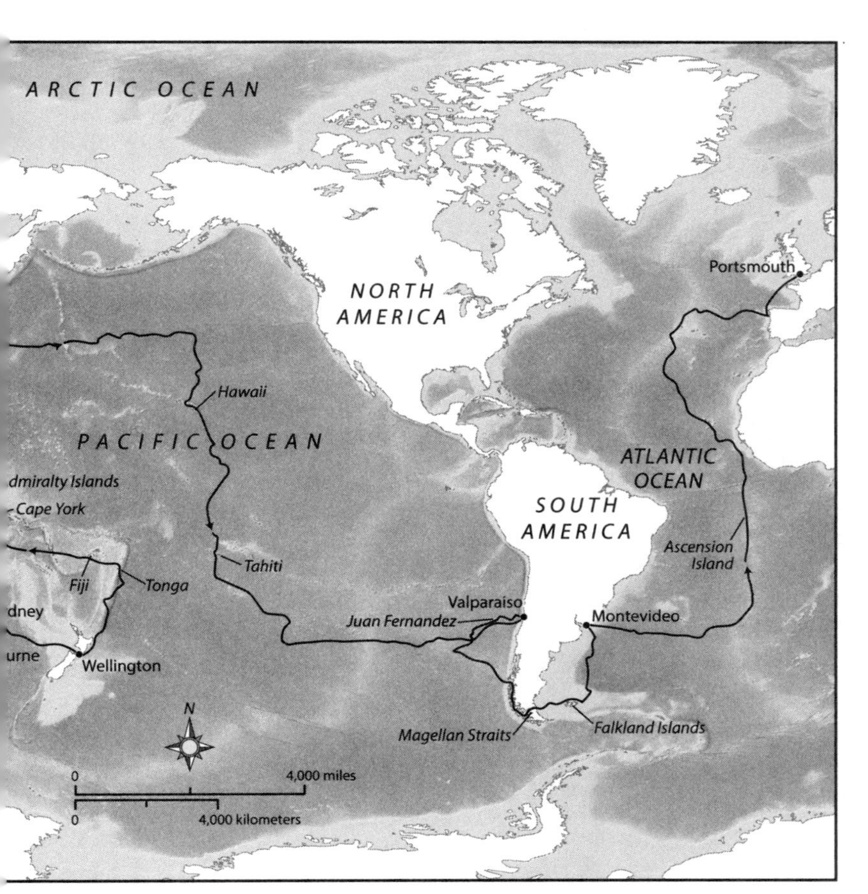

ARCTIC OCEAN
NORTH AMERICA
Portsmouth
Hawaii
PACIFIC OCEAN
ATLANTIC OCEAN
dmiralty Islands
Cape York
SOUTH AMERICA
Ascension Island
Tahiti
Fiji
Tonga
Valparaiso
dney
Juan Fernandez
Montevideo
urne
Wellington
N
Magellan Straits
Falkland Islands
0
4,000 miles
0
4,000 kilometers

PREFACE

Thomas Huxley—a powerful patron of the *Challenger* mission to explore the world's oceans in the 1870s—assured his Victorian audiences that "the most important sea fisheries . . . are inexhaustible." It's a striking reminder of the gulf that lies between nineteenth-century optimism and our current period of brutal stress for marine life worldwide. Huxley's complacency is a reminder too of the dangers of so-called shifting baseline syndrome. He believed in an inexhaustible deep sea because of its sheer vastness but also because its changes over time were invisible to him.[1]

Not so to us, thanks in great part to HMS *Challenger*, whose voyage 150 years ago helped elevate marine research from amateur shell collecting to a global science. *Challenger*'s naturalists mostly shared Huxley's naive view of the oceans. But their collective achievement—thousands of species collected, hundreds of reports written, reams of data points on ocean temperature and chemistry—have left a system-wide account of the 1870s marine world that is a critical baseline by which to measure oceanic change in the century and a half since. This book offers one set of such measurements: the Victorian ocean and our ocean, side-by-side, in sobering relief.

Precisely halfway between the legendary *Challenger* expedition and ourselves sits Rachel Carson, doyen of writers on the oceans. Her book *The Sea Around Us* (1951), published a

1 Huxley (1884) 14. Huxley's view represented a consensus among scientists and government officials on both sides of the Atlantic. Many working fishermen disagreed; see Bolster (2012).

decade before *Silent Spring*, was the first to describe marine life from the point of view of its creatures, with humans relegated to the margins as casual predators or bystanders. Carson's biocentric approach is my inspiration here, though I also introduce fresh archival material to this oft-told story of the sea. The journals of several of *Challenger*'s expeditioners that I rely on to evoke daily life aboard the ship—and dramatize key episodes of scientific discovery—have never before to my knowledge seen the light of day.

The Wake of HMS *Challenger*

INTRODUCTION

The Beach Philosophers

The experience of the pioneering research vessel HMS *Challenger* in the Southern Ocean in early 1874 summed up its contradictions. On the bleak Kerguelen Islands, just outside the Antarctic Circle, *Challenger* fell in with an American whaler crewed by desperate Portuguese teenagers who had escaped military conscription at home only to find themselves transported to a polar heart of darkness. Deserters from the whaling station were shot on the spot, while handwritten epitaphs marked the beach graves of those who had "drowned while fastened to a whale."[1]

Circumstances were more pitiful still for the marine wildlife of the Kerguelen Islands. The Americans had mastered the deadly art of firing bombs at surfacing whales from close range. When whales were scarce, the hunters slaughtered seals with no care for the sustainability of the colony. Breeding females and their pups were clubbed and skinned with the males. During their brief visit, the crew of the *Challenger* launched their own killing spree. In one encounter with a herd of sea elephants, they attacked the giant, floppy-nosed male with clubs and knives, then shot his cows as they fled into the water. The dying sea elephants, who had "a most enormous quantity of blood in them," stained the waters of the bay bright red.[2]

But in contrast to the whalers, where *Challenger* brought death she also discovered life—teeming, resilient, and wild.

1 Wild (1878) 65.
2 Moseley (1892) 177.

Boating in the cold waters off Kerguelen, *Challenger*'s naturalists navigated swarms of jellyfish and thick fields of kelp inhabited by never-before-seen mollusks and sea cucumbers. Their tow nets scooped up a smorgasbord of plankton—the bottom of the food chain. Then the *Challenger* trawl—designed for larger specimens—raised a stunning array of creatures from the frigid depths two miles down: gigantic glass sponges, tube-like sea squirts, skates, sea spiders, a type of worm called a sea mouse, frill-lined crabs that recalled the trilobites of the ancient ocean, and a bizarre crustacean with red-colored hexagons on its shell that looked uncannily like eyes. Seven hundred fifty species in all—not counting the plankton—two-thirds of them unique to the Southern Ocean. According to established Victorian science, life was not supposed to survive in the cold, dark abyss. But a dazzling menagerie of deep-sea animals flourished here in Antarctic waters, as they did wherever HMS *Challenger* explored the remote oceans across some 70,000 miles in the years 1872–1876.[3]

In the *Challenger* era, whaling and sealing figured among the handful of fisheries impacting the world's oceans at an industrial scale. Today, by contrast, overfishing is just one of a litany of devastating human impacts on marine life: from trawler deforestation of the seabed, to the bleaching of coral reefs, to flotillas of plastic scum twice the size of Texas. Carbon pollution is warming and acidifying the oceans, placing deadly stress on sea animals and their plant habitats that are adapted to narrow temperature windows. Since the 1980s alone, vertebrate ocean species—fish, turtles, dolphins, etc.—have declined by more than 20 percent, and the total fish population of the oceans by nearly 40 percent. *Challenger*'s meticulous inventory of deep-sea life in the 1870s thus represents the most

3 "Life is everywhere present on the sea-bed in all depths and at all distances from the shore." Murray (1896) 487.

FIGURE 0.2 HMS *Challenger* in Antarctica. January 1874. Attributed to William Frederick Mitchell (1880). Courtesy of Royal Museums Greenwich.

detailed picture we have by which to compare the oceans as they were in the first stages of human annexation.[4]

The *Challenger* story begins not with the ship itself—a Royal Navy corvette converted to a floating marine laboratory—but with the global ocean it navigated. This four-billion-year

4 See McCauley (2015). If the *Challenger* voyage has a precursor, it is the global expedition of 1845–1847 sponsored by the King of Denmark, with a suite of naturalists onboard. But due to the death of the king and political turmoil at home, the 93 containers of specimens collected by *Galathea* were not properly processed, and the results remained unpublished.

history, in turn, consists of dual, interwoven threads: a geological tale of slow-drifting continents and the gradual opening of modern ocean basins, in stark contrast to a biological record of boom and bust.

The year *Challenger* set sail on her epic voyage, reports reached Europe of the most ancient sea creature yet found—*Aspidella terranovica*. An amateur naturalist had stumbled on a constellation of conical rings imprinted on a rock in St. John's, Newfoundland. Whatever the enigmatic *Aspidella* fossil might be—a kind of jellyfish or a holdfast for a soft coral or sponge—it was the first evidence of the ocean's "sudden" evolution, 600 million years ago, from an intermittently freezing anoxic basin, host only to primitive microbes, to a complex aquatic environment capable of supporting prototypes of our modern corals, anemones, and crustaceans.

The so-called Cambrian explosion that followed turned the oceans into a global aquarium of wonders—including, for the first time, vertebrate fish. But continued volcanic upheavals—accompanied by splitting continents—rendered marine life vulnerable to rapid changes in oxygen and acidity. Fish populations swelled then collapsed. Even the ubiquitous crawling trilobites vanished. From the ruins of the worst crash of all—the end-Permian extinction 251 million years ago—modern calcite-shelled plankton emerged capable of transporting carbon from the surface to the seafloor. Ocean chemistry stabilized. For nearly 200 million years, warm water oysters thrived, while reptilian monsters dominated shallow, coral-filled seas that lapped deep into continental interiors. And so might the world have remained in essentials but for a rogue asteroid that brought an end to the ocean dinosaurs and greenhouse climate of the Mesozoic. In the aftermath of this latest brutal extinction, fish repopulated the cooling seas along with a cornucopia of invertebrate marine life: from tiny plankton of brilliant design to dazzling starfish to a deepwater squid ten meters long.

The genus *Homo* emerged 2.5 million years ago to a chilly Earth with its modern continents established and a colorful marine life recognizable to us today—at least in outline. It's worth reminding ourselves, however, that HMS *Challenger* did not set sail across an untouched Pleistocene ocean. Conventional histories of the first modern humans glamorize their decimation of land megafauna, but more recent archaeology spotlights our beachcombing instincts. Human communities began harvesting coastal marine animals at least 40,000 years ago. The first Americans, for example—who travelled the so-called kelp highway via California to Patagonia—feasted on abundant shellfish. Neolithic rock paintings from Korea depict the first known whale hunts.

Ancient shell middens, found along coasts and island beaches worldwide, are the clearest evidence of our long-term dependence on marine protein. As for the written historical record, measurable impacts of human fisheries on marine life trace back at least a thousand years. In the North Atlantic, depletion of freshwater fish stocks by the eleventh century turned attention seaward to the seemingly inexhaustible bounty of herring and cod. Europeans pursued these stocks across the Atlantic to Newfoundland then New England, where they fortuitously discovered oyster beds sufficient to sustain arriving New World migrants in their hundreds of thousands. The emerging global fishery—at the vanguard of European empires—plundered sea cows in the Bering Sea, turtles in the Caribbean, sea otters off California, and seals and whales to the furthest reaches of the oceans, including both poles. By the time *Challenger* set sail, hundreds of ships prowled the world's whaling grounds, while Scotland alone—intellectual hub of the expedition—was producing 800,000 barrels of herring annually for export consumption.

But rapacious as it was, the Atlantic fishery of the Victorian age paled in comparison to what has followed. *Challenger* herself was a hybrid vessel: part sailing ship, part

FIGURE 0.3 Centuries before the *Challenger* voyage, pursuit of the codfish drew European vessels to western Atlantic waters and the "new world." "Gadus Morhua, the Cod Fish" by Marcus Elieser Bloch (*Ichthyologie, ou Histoire Naturelle des Poissons*, vol. 1 [1796], plate 64).

steamship. With the launch in the late 1870s of the first modern steam trawlers capable of working day and night regardless of wind or tide, the fate of ocean ecosystems—our Pleistocene heritage—was sealed. Barely an inch of the world's continental shelves has escaped industrial trawling in the century and a half since, transforming millions of acres of teeming, plant-rich seabed into sandy underwater desert. For millennia, coastal human communities confined themselves to fishing well-stocked offshore waters. In the course of a short century, we have become the oceans' apex predator, eliminating an estimated 90 percent of fish in the upper levels of the food chain. By historical chance, *Challenger* set out on her global survey just prior to the industrial steam era. The naturalists who sailed with her have left us a vivid account of that marine world now lost to us.[5]

Our sense of *Challenger*'s unique value is the more acute when we realize that so ambitious a project would never have happened but for a momentary confluence of interests within a well-resourced maritime empire. The British, of course, were

5 Myers and Worm (2013).

not the first to seek to rule the waves. The islands of the world's largest ocean—the Pacific—were navigated and settled by seafaring Lapitans three thousand years ago. By the common era, the Indian Ocean played host to busy commercial traffic connecting ports in India, Africa, and Europe. Chinese traders, meanwhile, extended the "silk road of the sea" from the Southeast Asian archipelago via the Red Sea to Mediterranean markets, and Viking traders (and raiders) dominated the North Seas, venturing across the Atlantic Ocean to the Grand Banks.

These regional maritime powers flourished for centuries at a time. But no concept of a global ocean existed until the 1500s when Dutch and Portuguese navigators rounded the perilous southern capes connecting the Atlantic to the rest. Fast forward to the 1870s, and the global infrastructure of the British seagoing empire—with its well-supplied ports on major coasts in all oceans—combined with the emergence of a professional science community capable of effectively lobbying government, launched *Challenger* on her multiyear research mission. It was an unprecedented circumstance in maritime history, destined never to be repeated.

This book offers an inventory of *Challenger*'s 1870s global ocean but also of human communities under the European colonial yoke. The anonymous dockhands who filled *Challenger*'s hold with coal, and helped pilot the ship across treacherous reefs, belonged to the ranks of displaced and enslaved peoples whose labor fueled the engine of empire for centuries after Columbus. In addition to these forced human migrations, we will get a view of the coastal and island environments already ravaged in the 1870s by invasive terrestrial species, plant and animal—these were the rogue camp followers of European expansion across the globe in the nineteenth century. By activating our time-lapse imaginations, we will see how the fate of the mostly pristine Victorian ocean was anticipated in the damaged landscapes *Challenger* encountered.

Deforested coastlines in Asia and the Americas, emptied of their wild animals and long-term resident peoples, are today mirrored in devastated reef systems and a depopulated deep sea.

While Victorian colonial geography is the setting for this book, colonialism is not its subject. Marine life and its ocean habitats are the subjects. What, then, was the state of marine science in 1872, when the well-resourced *Challenger* naturalists set sail to plumb the oceans' unexplored depths? For the first half of the nineteenth century, the study of the oceans and its creatures was mostly shorebound and decidedly amateur. Like most Victorian scientific pursuits, it attracted miscellaneous intellectuals, clergymen, and eccentrics. Because empirical data on the sea was lacking, early marine science relied heavily on speculation: it was beach philosophy and little more.

By the 1870s, however, advances had been made. The coasts of Europe—and the Eastern Seaboard of the United States—had been energetically trawled for specimens. American Matthew Maury had published a compendium guide to ocean winds and currents based on the accumulated data of old ships' logs. And the richly various marine tribes—called phyla—had been organized into something like their modern form. Charles Darwin's publication of *The Origin of Species* in 1859, meanwhile, had given special impetus for exploration beyond coastal shores. The deep sea, it was speculated, might harbor missing links between extinct life preserved in fossils and contemporary plants and animals.

A generation prior to *Challenger*'s voyage, the charismatic Manx naturalist Edward Forbes had retraced Aristotle's footsteps on the shores of the Aegean, where he reflected on how little Western marine science had advanced since the Greeks. On the beaches of Lycia, he watched local fishermen hunt for cephalopods by torchlight, paddling stealthily in the rocky shallows armed with spears. From a borrowed fishing boat, he observed the sea "filled with glancing needles of glass." The mollusk *Criseis*, with feet like butterfly wings, danced across the water, catching the sunlight on its transparent, pointed shell. Further out to sea, Forbes saw the sponge divers peering from their boats into the sunlit water just as in Aristotle's day. Holding their breath for minutes at a time, they brought up squishy masses by the basketful.[6]

For the first generation of oceanographers, data-gathering consisted of lowering a crude trawl from the stern of a fishing boat and allowing it to drag at random along the seabed—so-called dredging. From his pioneering research in the Aegean Sea, however, Forbes drew a fatefully erroneous conclusion. The mollusks, in their wild variety, studded the Aegean seafloor from the shallows to depths of several hundred feet, feeding among the seaweeds and corals. As depth increased, their shells changed: the vivid, patterned colors of the shore varieties faded to a bland whiteness. Then, beyond a critical limit—about a quarter-mile deep—the mollusk population declined to near zero, prompting Forbes to speculate that marine life was impossible in the dark, cold depths of the sea.

Challenger would definitively refute Forbes's "azoic" theory, but his ecological approach revolutionized Victorian marine science. In Forbes's official report on the Mollusca of the Aegean Sea (1843)—a publication that *Challenger*'s scientific director, Charles Wyville Thomson, hailed as "an era in the progress of human thought"—he argued that the distribution of marine life was determined by three factors: climate, water chemistry, and depth. Ocean depth, in turn, could be organized according to discrete biotic zones extending from the shore to the abyssal plain, fauna and flora changing as it went. Forbes died in 1854 having just acceded to the chair of natural history at the University of Edinburgh, a position Thomson now held. With an eye to Forbes's legacy, *Challenger*'s

6 Forbes (1843) 102.

awesome task was to extend his beach philosophy to the remotest stretches of the global ocean.[7]

Departing the Kerguelen Islands in February 1874, the *Challenger* crossed the Antarctic Circle, continuing to dredge the teeming polar deep while dodging icebergs. Tacking southward in the ice pack, the ship was the furthest distance from home in its yearslong voyage of discovery—8,500 miles. As if to signal that some physical limit had been reached, the on-deck barometer sank to a sickening low. The biting south wind off the ice freshened to a gale.

Challenger's veteran captain, George Nares, had seen enough. He had strict orders not to penetrate the ice pack and had only three months' provisions were they to get stuck—insufficient to survive the winter. The *Challenger* crew, huddled below deck, heard the long-awaited order to set sail for Melbourne, their safe haven two thousand miles to the northeast. On deck, the whipping snow stung their faces. The captain had sought shelter in the lee of an enormous iceberg, but a shift in the wind drove the ship directly toward it. Shouted orders were muffled by the howling gale, so the men simply abandoned their posts, sliding down the ice-slickened ropes to the deck. They expected any moment to hear the mainmast crashing down around them. That wild day in the Antarctic ice pack, the fate of the first global exploration of the deep sea hung in the balance.

In any ordinary naval vessel, imminent collision with an iceberg—in gale conditions with near-zero visibility—would call for all hands on deck. But on HMS *Challenger*, six able-bodied men remained below during the crisis, practically useless. John Murray—a junior scientist from Edinburgh in charge of

7 Thomson (1873a) 266. Thomson was known personally and professionally as Wyville Thomson until his knighthood in 1876, when he became Sir Charles Wyville Thomson. For this narrative, I have preferred the name by which he was known aboard *Challenger*.

skimming the sea surface with nets—complained of rheumatism in his diary while listening apprehensively to the "excitement" on deck. Ship's artist, Jean Jacques Wild, feared the worst: his leather folder full of illustrations of deep-sea wonders would surely sink without trace, along with their bottled originals. Meanwhile, the young German naturalist Rudolf von Willemoes-Suhm (whose destiny was to die in a different ocean) felt only frustration that bad weather was keeping him from his dissections in *Challenger*'s state-of-the-art workroom. As for their three senior colleagues—Wyville Thomson, Henry Moseley, and John Buchanan—one, Moseley, took to his bed with symptoms of exhaustion. But none recorded his feelings the day *Challenger* crashed the iceberg, which shattered her jib boom, sent a man plummeting from aloft, and flung a ship's boy over the side (miraculously rescued: half frozen).

In the eyes of *Challenger*'s men, these lubberly additions to the crew—nicknamed "the scientifics" or "philosophers"—were "as unlucky a shipmate as a cat or a corpse." It was because of the philosophers that *Challenger* was in Antarctica at all, having already crisscrossed the Atlantic Ocean four times in the previous year before rounding the dangerous Cape of Good Hope. Still ahead of them lay the vast coral waters of Asia and the Pacific and two more years scouring the ocean floor under all conditions, day after tedious day. Even to the *Challenger*'s officers, men of some education, their mission often seemed little more than to indulge the naturalists' unaccountable taste for wriggling critters in freezing slime.[8]

To us, however, the *Challenger* expedition, on its 150th anniversary, represents a unique nineteenth-century encounter with the sea at a planetary scale—a scale we take for granted in our era of global observation systems, deep-sea submersibles, and

8 The observation about oceangoing scientists and bad luck is from naturalist Joseph Hooker, who sailed with the Antarctic exploring expedition led by James Clark Ross (1839–43). Hooker (1877) 351.

data-rich ocean models. Without *Challenger*, we would have little notion of how much of the Victorian marine world we have since lost to overfishing, pollution, and climate change—and how much we are still in danger of losing. A century and a half after *Challenger*'s spectacular odyssey—which returned thousands of marine specimens new to science and collected groundbreaking ocean temperature data—it's worth dwelling on the sheer unlikelihood of the British government, in 1872, agreeing to send a warlike vessel, complete with seagoing philosophers, to the distant reaches of the world, far beyond all trade routes and conventional horizons. To this day, some remote stations where *Challenger* hauled up a museum's worth of spectacular sponges, starfish, and other salty wonders have never been revisited.

For a half-century before the *Challenger*'s epic voyage, a loose network of marine naturalists in Europe and the United States had been mostly shore bound, rarely venturing beyond knee-deep in the shallows, or at most a quarter mile from shore to dredge up mussels and sea stars in co-opted fishing boats. Telegraph cables—a new deep-sea technology—had awakened official interest in the ocean floor on both sides of the Atlantic Ocean. But it was a more nebulous set of circumstances—the Victorian fashion for all things marine combined with the chance influence of a few well-connected scientists—that saw study of the oceans suddenly promoted, in the imperial form of HMS *Challenger*, from shell-collecting and amateur dredging to prestige science on the global stage. *Challenger*'s cutting-edge mission would test not only how far the seafaring imagination could extend but also how deep.

The voyage of HMS *Challenger* has been extensively chronicled by ocean historians to the point of becoming a crumbling Victorian monument all its own. My new approach to the famous journey has been to de-emphasize its human actors—all white, male emissaries of empire—in favor of the global ocean itself and its creatures: those wild natives of the global aquarium the Victorian naturalists set out to explore, and which today face mounting pressures from a rapidly deteriorating marine environment. That said, to do justice to the entire inventory of *Challenger*'s discoveries—or even a significant fraction—is a gargantuan prospect that drove its director Wyville Thomson to an early grave. The sheer volume of *Challenger*'s scientific legacy has also worked to diminish the expedition's appeal in the popular imagination.

To forestall both outcomes for this book, I have chosen a select menagerie of marine animals to tell *Challenger*'s story, each creature representing a leg of the epic journey. Interspersed with these, I recreate the major *physical* oceanographic discoveries of the voyage: these include the circulation of globe-girdling ocean currents; manganese nodules in their trillions on the Pacific Ocean floor; and the famed Challenger Deep, five miles beneath the pearl-blue surface, near Guam. I will describe, too, the remarkable phenotypic traits common to ocean organisms—such as bioluminescence, camouflage, and the daily mass vertical migration from the deep sea to the surface—that fascinated the *Challenger* philosophers, and now represent full-fledged subfields of ocean science. This book spans the nineteenth-century aqueous globe, but it also travels across time. The biography of each of my *Challenger* animals, for example, includes its evolutionary past and present existence, as well as its future prospects in a radically changing global ocean.

The oceans, it is said, are the last wilderness, and they are under existential threat. This book offers a last-chance tour of their wonders. Marine life here is the truly Other: the nonhuman in its extravagant, uncaring glory. Its gloomy habitat, the deep sea, is a wilder place than any terrestrial forest or remote highland. From a self-amputating sea star, to the endangered green turtle, to the ever-present animalcules glowing by their millions in *Challenger*'s frothy wake, the creatures of the ocean deep make a sometimes-elegiac transit across these pages. In classic histories of oceanography, the *Challenger* expedition marks the birth of an era. But *Challenger* also signifies the end of innocence: its intrepid voyagers bore witness to the last days of the preindustrial ocean. One hundred fifty years on, we live in its wake.

But we would be wrong to project our present anxieties onto the past. The Challengers—the ship's officers, crew, and naturalists—felt no sense of loss on their journey, let alone the psychological cloud of a planetary emergency. Rather, the oceans supplied "endless novelties of extraordinary interest" to be pursued to the literal ends of the Earth. Across 1,250 days and 70,000 miles, their nets and trawls bulged with unseen marvels from the deep sea. On their return, it took twenty years to process them all and publish the results.[9]

The ocean has a different appearance and history depending on who is looking at it. The *Challenger* naturalists contemplated its mysteries through Western eyes and used ready-made tools of European scientific inquiry to describe what they saw. Some of that language is obsolete or discredited—but core insights from the mission have endured, which I detail in the chapters to follow.

I also have an argument to make. Implicit in the sprawling *Challenger* canon of early ocean research is a core environmentalist principle: that to preserve Earth's animals and habitats requires awakening our dormant biophilia—our love of nature. To save the world, we must re-enchant the world. The *Challenger* scientists' attitudes toward nature were different from ours—underwritten as they were by a belief in European superiority. But this is not the whole of their legacy. Wyville Thomson's and his colleagues' deep passion for the life aquatic is palpable on every page of their voluminous writ-

9 Thomson (1873a) 49. Imperial measures are appropriate to the period of HMS *Challenger*. Metric units will be cited from current scientific literature.

ings. Their biophilia, when separable from their prejudices, stands as a model and inspiration. In this book we will peer over the shoulders of these nature-loving Victorians at work and, through their eyes, bear witness to the undersea marvels they encountered.

Like a telescope whose sweep of the empty horizon catches a distant blur then zooms in until a single, white-sailed ship fills its entire frame, the narrative that follows tracks *Challenger* in close-up detail on her ocean-spanning travels of 1872–1876. To best capture the still urgent relevance of that journey to us, this book operates as a work of historical ecology: reconstructing the past ocean to better understand its current deteriorated state. To armchair travelers, the book offers too a unique postcard portrait of the late nineteenth-century world, when European nations maintained and extended their power through their navies and colonial settlements. As an adapted warship herself, HMS *Challenger* belonged to that colonial mission, while also pointing to an ideal that lay beyond crude imperial objectives. The goal of the *Challenger* expedition was nothing less than the first global census of deep-sea marine life.

Considered from a planetary viewpoint, my thirteen ways of looking at HMS *Challenger* add up to no more than a snapshot from the oceans' unimaginably long history. The *Challenger* expedition was, after all, just one sea voyage among millions. But it was the most ambitious, well-documented voyage ever dedicated to study of the sea. One hundred and fifty years on, HMS *Challenger*, of all ships that sailed, still has the most to tell us about our indispensable oceans—past, present, and future.

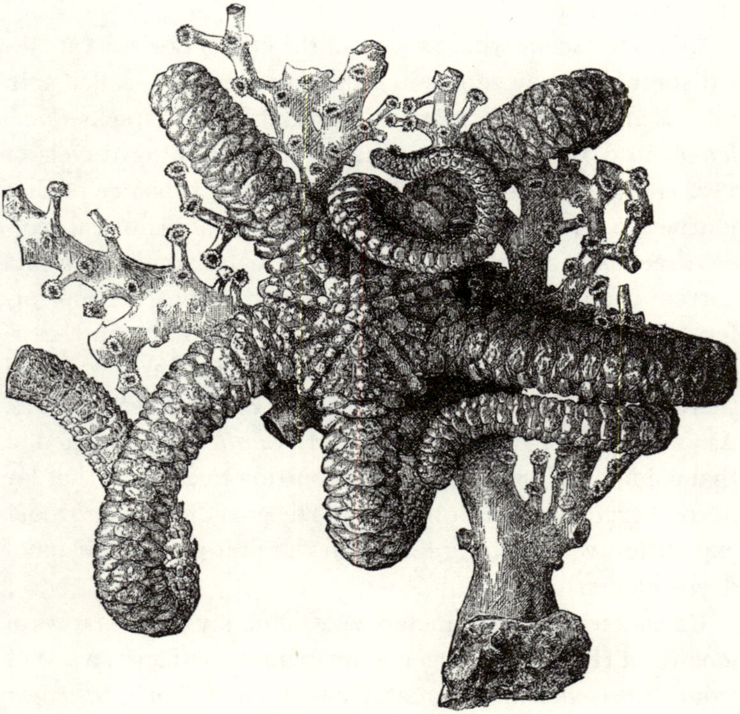

FIGURE 1.1 Brittle star *Ophiomusium lymani*. From *Bulletin of the Museum of Comparative Zoology* (1863), p. 175. (Inset) HMS *Challenger* Leg 1. Portsmouth to Gibraltar. December 1872–January 1873.

CHAPTER 1

Embrittled Star

Cape Espichel, Portugal
38°24′ N, 9°13′ W

Edward Forbes—father of the Victorian beach philosophers—spent much of his short, briny life wading among the coastal rock pools and shallows of the British Isles. Along the scallop banks, between tides, he often came across the common brittle star writhing about in numbers. Laying hold of a snake-like arm, Forbes watched, mesmerized, as the creature broke into pieces, shedding its limbs with shocking abandon. The self-amputated arms then disintegrated into even smaller bits until he was holding only the flat, disc-like body of the sea star amid the jumbled ruins of its former being. "Touch it," Forbes wrote, "and it flings away an arm; hold it, and in a moment not an arm remains attached to the body." The only recourse for the naturalist—to preserve this wild, puzzle-piece creature for his collection—was to plunge the star in a tub of cold fresh water, paralyzing it, and so preempt its spectacular self-demolition.

The early Victorian generation of marine naturalists could not confine their curiosity to the beach and took to primitive dredging from fishing boats. There, in deeper waters, the brittle star proved ubiquitous. In his countless offshore ramblings in the 1830s and 1840s—off misty beaches in Scotland or Wales, or his native Isle of Man—Forbes's dredge emerged filled with undulating ophiuroids.

When the local oystermen he had hired emptied the sodden dredge bag on deck, the "serpent stars" rioted in all directions, spraying limbs. These autotomic projectiles weren't inanimate, either: the discarded members contorted themselves in a grotesque tribute to their mother form. Forbes theorized the creature self-destructed to deter its predators, which was certainly the effect on his dredging companions. Thoroughly spooked, the locals begged Forbes to allow them to shovel the wriggling parts back into the sea. He refused, naturally. Under the microscope, a single fragment of a brittle star arm exhibited an architecture of rare "lightness and beauty," worthy of a cathedral spire.[1]

Forbes never witnessed a brittle star regenerate it lost limbs. Perhaps he never imagined the possibility. In fact, strategic embrittlement is common among marine creatures, who enjoy regenerative powers that terrestrial mammals have apparently lost. What the Hebridean boatmen assumed was a traumatic, unnatural scene of self-amputation among the brittle stars was in fact ordinary autotomy—a survival instinct—and a natural prelude to regeneration (albeit not in that instance).

The arm of an ophiuroid has inbuilt points of weakness at the skin, ligament, and nerve, allowing for a clean break in moments of stress. The collagenous arm tissue—light on cells—minimizes damage to the severed limb, allowing regeneration to begin immediately. Weight-bearing terrestrial mammals mostly dispensed with regeneration—what to eat while the limb grows back?—but in the buoyant seas, a legion of loose-limbed creatures from crabs to octopuses to sea stars will simply reproduce what has been lost, like a rebooted embryo. In this way, a brittle star might sacrifice an arm to a passing cod, crustacean, or naturalist, and live to tell the tale.

The name ophiuroid is derived from the Greek *ophis*, meaning snake, and *oura*, tail. The class comprises more than two thousand species, found in all oceans from the intertidal beds of seashores to the remote depths. For modern biologists, the

1 Forbes (1841) 63–64.

ubiquity of ophiuroids—and their worldwide preservation in the fossil record—holds a key to our understanding of marine biodiversity over evolutionary time. But at the dawn of ocean science, 150 years ago, the beautiful weirdness of the brittle star was emblematic of the still-unexplored deep seas.

For eminent Victorian Wyville Thomson—one-time dredging companion of Edward Forbes and chief scientist appointed to HMS *Challenger*—the wriggling ophiuroids with their five-point, crystalline arms raised core questions the expedition had been charged to investigate. How far into the open sea did the domain of shoreline creatures like the brittle star extend? Would the ocean depths offer up an entirely different fauna—a vast museum of marine life where the monsters of the fossil record lived on? Or would the brittle star be just as populous in the deep sea and prove itself, and its circus-trick of autotomy, a grand success story of aquatic nature?

The first month of HMS *Challenger*'s three-and-a-half-year odyssey involved a welcome exchange of one climate for another: stormy English winter for Iberian spring. Even under full steam, the ship had made rough headway in wild December weather out of the southwest, taking four days to run off the hundred miles of English coast from Sheerness to Portsmouth. In the dead of night, huge seas flooded the engine room, crushed a life boat to pieces, and carried away the jib boom with all sails. Dawn brought news that several ships plying the same course had been wrecked.

The *Challenger* had already lost one crew member, who fell from the gangway at the Sheerness dock. A diver had located the dead man the next day sitting up in twenty feet of water, his face erased beyond recognition. In these bleak first days of *Challenger*'s epic voyage, whispers of an unlucky ship doubtless passed, unrecorded, before the mast, but have left no imprint on the dry prose of the ship's log or the officers' published recollections.

For Thomson and his fellow unseaworthy scientists on board, their first impressions of life onboard *Challenger* were uniformly negative. Barely able to distinguish night from day, the philosophers crawled about their cabins, revoltingly ill, and captive to a chorus of unearthly sounds: groaning timbers, the howl of the wind in the rigging, the captain's hoarse voice struggling to be heard, and the tramp-tramp of his harassed crew a few feet overhead. The ship's dogs, all in a panic, yelped through the endless night. Sleep deprivation and all-round misery did not exempt them from what was arguably the worst of it: enforced jollity in the officers' mess. Appalled at this treatment, the *Challenger*'s scientific staff abandoned ship at Deal and completed the passage to Portsmouth via the train.

Christmas out of Portsmouth brought no respite, however. No sooner had the English shore faded into the twilight of the year's shortest day than *Challenger* began to pitch and roll into a relentless southerly on the Channel. The yuletide meal was spent with knuckles gripped on furniture and sweeping up smashed crockery. Thieves from the lower deck took advantage of the confusion to filch a turkey and roast goose from the officer's galley. Bone fragments were discovered in the maintop, but the investigation was not pursued. *Challenger*'s was not a naval mission. The men would be made to perform tedious tasks unfamiliar to them for the next four years, all while subsisting on starvation rations of salt pork, pea soup, and biscuit. Captain Nares and Wyville Thomson were invested with god-like authority on board to punish infractions but valued more the sailors' long-term goodwill.

Farther south, the Bay of Biscay, true to its bad reputation, presented a continuing chaos of wind and waves. When the pale disc of the sun finally appeared from behind the clouds, it revealed a mass of bobbing oranges surrounding a ship's hull, bottom up—a doomed cargo out of Portugal. Captain Nares declined to lower a search boat in the foaming sea. Even once arrived in the safe haven of Lisbon—six miles from the coast along the River Tagus—the foul weather did not relent. Adding to the frustration of a sluggish two weeks' crossing from Portsmouth, the Challengers spent a further eight days languishing at anchor in Lisbon waiting for the wind to shift.

So far, the expedition's main business for which it had been expensively crewed and provisioned—dredging the ocean floor for novel life-forms—had been ventured only three times. On the second day of 1873, off the northwest coast of Portugal, they had lost 1,700 fathoms of sounding line over the rolling side with valuable thermometers attached. The *Challenger*'s much-vaunted dredge then completed its debut upside down tangled in rope. The entire ship's company crowded the main deck to witness the awkward contraption disgorge its deep-sea mysteries—which consisted of a quantity of slime, a few dismembered starfish, and a solitary shrimp. On the next try this dredge, too, vanished along with two sounding lines.

The citrus perfume of the orchards of Lisbon told them they had left England several degrees to the north. While the battered ship took on coal for their Atlantic passage, the *Challenger*'s officers and scientists, with journals in hand, contemplated the architectural wonders of Lisbon. The magnificent Jerónimos Monastery, built on the spot where Vasco de Gama embarked for the Indies in 1497, boasted original cloisters to lure pilgrims of art. The sinuous lines of the arches flowed upward to the niche like a coastline merging into the sea. Later, Professor José Vicente Barbosa du Bocage led his distinguished English visitors on a tour of Lisbon's new zoological museum. There they wandered among cases of seashells while sharks' heads stared at them from the walls alongside a ghost-like chimera and a pair of stuffed manatees.

At the busy market, the local women unloaded baskets of fresh-caught fish while the men stood by smoking and chatting. The Challengers inspected their streetside meal with a professional eye before devouring it—the fried mackerel was oily and delicious—then gorged themselves on Mediterranean fruit. According to local courtship customs, young men were not admitted to the houses of unmarried women. So, as the evening drew on, the officers of HMS *Challenger* entertained themselves eavesdropping on public interviews between suitors in the street and their girlfriends leaning on the balconies above. It being the season for masquerades, dominoes passed them on the streets, capes flying. The church bells played popular arias every evening at six. Unable to resist this siren call, the *Challenger* philosophers nightly made their way to the opera house.

When, at last, the winds abated, *Challenger* retraced her course along the River Tagus toward the Atlantic swell. From the river mouth, Lisbon appeared in an entirely new aspect. Warm tints of color lit the city steeples in perfect imitation of an English summer's evening and seemed to promise a change in luck with this balmy change of latitude. Sure enough, the week's long course to Gibraltar that followed brought blue skies free of fog, a steady breeze, and a new determination to master the complicated art of dredging for deep-sea marine life.

On the bright morning of January 13, 1873, the Challengers watched the sandy cliffs of Cape Espichel fade from view. Twenty miles offshore, Captain Nares shortened sail and turned the ship's head to the breeze, bringing *Challenger* to a lilting halt. The hands mounted the heavy block on her main yard, then threaded the lead line to the deck, secured by a hook. The small port side deck engine the men disparaged as "the donkey" rattled to life. A light splash signalled the entry of the sounding rod over *Challenger*'s side into the sea, with iron weights to speed its journey to the bottom and thermometers to collect temperature readings en route. Giant rubber bands stretched from the block aloft to the deck as insurance against disaster, to relieve the straining lead line when, inevitably, the ship's wandering motion dragged it along the seafloor and it became caught on some unseen reef or obstacle.

Sounding the ocean floor while measuring water temperature at successive intervals occupied the ship's company until the early afternoon. In addition to thermometers, a flask was attached to the sounding wire for the collection of water samples, and yet another device for gathering sediment from the bottom. Once this first phase of data collection was complete, the officers, led by navigator Thomas Tizard, replaced the sounding rod with the dredge for collecting more sediment and any unsuspecting creatures from the ocean floor.[2]

Long, patient hours later, the dredge's great bag, attached to an iron frame, loomed beneath the hull like a breaching whale, bulging (they hoped) with deep-sea novelties. The lead line creaked with the strain, and the donkey blustered a half-octave higher. The *Challenger* scientists in their blue serge jackets leaned over the ship's side. They scanned the dimming horizon, hoping they would not lose all light by which to dissect, draw, and notate the contents of the dredge before it rotted and the long day's effort be wasted. Their greatest anxiety came as the unwieldy bag struggled to break free of the clinging meniscus of the surface.

In the long months and years ahead, the *Challenger*'s crew would tire of the spectacle of the suited scientific gentleman gathered around the large tub on their knees, sleeves rolled, clawing through the cold grey mud brought from the ocean bottom like children at a birthday treasure hunt. But for now, they savored its ridiculousness. The sight of the captain's nine-year-old son splashing about unrebuked among the philosophers made it a scene worthy of *Punch*.

For the scientists themselves, the serious business of HMS *Challenger* to which they had agreed to sacrifice four years of their lives had begun at last. This first successful dredge,

2 Tizard served a critical role as liaison between the naval officers and scientific staff aboard *Challenger*. He was later recognized with shared author credit for the expedition narrative.

FIGURE 1.2 Examining the contents of the dredge. Frederick Whymper, *The Sea: Its Stirring Story of Adventure, Peril and Heroism* (Cassell, 1887).

off the ocean mouth of the River Tagus to a depth of 470 fathoms (860 meters), raised a slithery host of worms, mollusks, sponges, starfish, and spiky sea urchins. With his first glance at their haul, Wyville Thomson spotted shell specimens he had seen during his prior dredging expeditions in British waters: the pearl-colored shell fringed with tiny filaments of the *Limopsis minuta*; the minaret-styled gastropod *Amphissa acutecostata*; and *Dacrydium vitreum* with its opalescent shine.

Rudolf von Willemoes-Suhm, meanwhile—the young German morphologist recruited at the last minute to *Challenger*'s scientific corps—was delighted to recognize half a dozen worm species from his own northern collections. It was the first clue, brought home to them during the subsequent four years, that life in the deep sea was profoundly cosmopolitan. Successful species, over immeasurable stretches of time, had reproduced, adapted, and colonized vast swaths of the ocean floor.[3]

The naturalists of HMS *Challenger* now faced a race against time to transport their precious specimens in buckets to the specially fitted laboratories on the main deck below. There they scrutinized each shell, worm, and squirming larva under the microscope. Peering into the lens, they scribbled notes with their free hand, produced a hurried sketch, then deposited the animals in spirit jars to be labelled and stowed away—all before the inevitable rot set in. Willemoes-Suhm worked closely with his senior colleagues in the *Challenger* laboratory—the ever-enthusiastic Henry Moseley and John Murray—to inventory the bulk of the catch, while Professor Thomson took choice selections for close study in his private salon. These included a large brittle star with snake-like arms

3 The long-accepted view of low biodiversity in the deep sea has been challenged in recent decades by studies of animal remains sampled from sediment cores and by a new evolutionary paradigm—cryptic speciation—whereby morphologically identical organisms may belong to different species. By this alternative accounting, the numbers of deep sea species potentially inflate into the millions; see Fišer (2017).

that Thomson had already had the good luck to discover and name: *Ophiomusium lymani.*

The brittle star sat limp in a pool of lamplight, its mystical allotment of five arms in various states of intactness. Thomson inspected the familiar diamond-shaped plates on its small, circular body and found himself struck again with its porcelain perfection. As with all his significant finds, Thomson's thoughts turned to whether its ancestors—identical or nearly so—might be found in fossil form and so advance his favorite thesis: that the deep sea was a museum stocked with ancient life forms long vanished from the Earth's surface. Evolutionary questions aside, it required a visit downstairs to the well-stocked *Challenger* library (which doubled as a card-playing room) to educate himself on *O. lymani*'s modern history, to which the expedition would now add a chapter authored in due course (he fully intended) by the American who bore its name.

Theodore Lyman—Victorian guru of the brittle star—had passed the years prior to the Civil War buried in the rich invertebrate collections at Harvard University. Since the arrival from Europe of his charismatic mentor, Louis Agassiz, amateur beachcombers from New England to Florida had submitted their prizes in watertight packages to Cambridge, Massachusetts. While the in-demand Agassiz proselytized for public science across the country, Lyman, still in his early twenties, immersed himself in the mysteries of ophiuroids. When he had exhausted the collections at Harvard, he travelled to Washington, DC, to study the repositories of the United States Coastal Survey, which, since 1844, had been ordered to preserve all specimens brought up with soundings along the nation's coasts. Lyman's taxonomist's eye relished the sight of hundreds of brittle stars suspended in jars, labelled with date, depth, and location, and arranged neatly in cabinets at the Office of Coast Survey on New Jersey Avenue.

Lyman's life changed on a research trip to the Florida Keys, sent by Agassiz to test the proposition that marine life was horizontally distributed according to temperature, just as Alexander von Humboldt had identified the vertical organization of terrestrial fauna based on altitude. The Florida Keys in the 1850s, however rich in marine life, was sorely lacking in human accommodation, so Lyman considered himself lucky to meet with a team of army engineers constructing lighthouses. Their commander was a gruff but hospitable man named George Meade, who took a liking to the student of the renowned Agassiz. When, in the bloody month of September 1863, Meade ascended to the command of the Army of the Potomac, he remembered his young biologist friend and appointed him to his staff. Lyman, temporarily sidetracked from his career as a marine biologist, carried the flag of truce to Rebel lines after the massacre at Cold Harbor and shook Robert E. Lee's hand at Appomattox.

After the war, it was the turn of another Agassiz student—a young, aristocrat émigré named Count de Pourtalès—to return to Caribbean waters, this time aboard a vessel equipped with a dredge. The Atlantic beaches had been well scoured, but the deeper seas offshore were yet a mystery. Did the shallow-water species vanish at depth, or transmute, or give way to an entirely original fauna? Yellow fever brought his first research mission to a quick, fatal end, but the Count, undeterred, returned in 1869 to dredge the waters off Cuba, from where he delivered a fresh trove of floating brittle stars to Theodore Lyman's laboratory at Harvard. With that single shipment, the world catalogue of known ophiuroids more than doubled, and a dividing line separating littoral and deep-sea marine life first came, albeit faintly, into focus.

Casting his eye across the samples sent by the Count de Pourtalès, Lyman had instantly recognized the beach specimens—the chocolate brittle star and spiny brittle star—from his own wadings among the corals and sponges of the Florida shallows.

These were consistent with what he understood about the emerging law of latitudinal zones: the sea stars of Cape Cod were as distinct from those of the Carolinas as the Caribbean ophiuroids were from specimens from the beaches of Brazil further south. Moreover, what de Pourtalès had dredged up in depths a moderate distance from shore—between 15 and 75 fathoms—seemed lightly modified versions of these shallow-water species. But from those jars marked with depths greater than a hundred fathoms—well out to sea—an entirely new brittle star fauna emerged. "New," at least, to human eyes. One original genus, which Lyman had christened *Ophiomusium*, with its porcelain sheen and spiny arms like cut glass, seemed to recall extinct sea stars excavated from the fossil beds of Europe.

For the naturalists of HMS *Challenger*, Lyman's deep-sea *Ophiomusium* was an entirely novel brittle star that did not conform to the zonal rules governing its shoreline brethren. Interest escalated with a companion discovery on the far side of the Atlantic. Barely had Lyman written up his report on the remarkable discoveries of the de Pourtalès expedition than Wyville Thomson himself, dredging off the coast of Ireland in the summer of 1869 aboard HMS *Porcupine*, raised a second allied species of *Ophiomusium*, which he named *O. lymani* in recognition of his American colleague's yearslong labors.

Near Rockall Trough off the Irish coast, brittle stars and their jettisoned parts, which had adhered themselves to the sounding line, literally rained down on the *Porcupine*'s deck in a shower of light. In some places where Thomson had ordered the dredge, almost everything they raised—from ribbons of plankton, to corals, starfish, and sea worms—shone with phosphorescent brilliance. Even the cold, grey mud dirtying the deck lit up with flashing diamond specks. At night, he could read his watch by the lambent light of the soft corals and sea pens. It was here, amid "an extraordinary abundance of animal life," that Thomson first laid eyes on an Irish *Ophiomusium*

FIGURE 1.3 A brittle star fossil (*Ophiopetra lithographica*) found near Regensburg, Germany. Fossil from the Lower Hienheim Beds (Lower Tithonian, Upper Jurassic) (Wilson44691 / Wikimedia Commons CC0 1.0).

lymani. Given the creature was snatched from the ocean floor half a mile beneath the rocking hull of the *Porcupine*, he was almost certainly the first person to do so.[4]

Brittle star fragments from the time of the dinosaurs were known to Thomson's predecessor Edward Forbes. Railway diggings in the clays of London and Oxford had exposed ancient amputated ophiuroid limbs, time's messengers from when those renowned cities were shallow, coastal marinas innocent of humanity—and of trains. Even in the 1840s, Forbes was not the first to remark how essentially similar these ophiuroids—millions of years old—appeared to their modern kin. The oldest brittle star fossil predates the end-Permian extinction 251 million years ago, which obliterated 90 percent of all marine species. Modern deep-sea explorers—cruising off the Florida coast in Theodore Lyman's wake—have recovered *Ophiomusium*-like fossils dating to the Early Cretaceous, 114 million years ago,

4 Thomson (1873a) 98–99. "Phosphorescence," which the Victorians defined as to glow without burning, is now an obsolete term for bioluminescent marine life.

suggesting that brittle stars have nonchalantly survived what were thought to be wholesale ocean extinction events, including the asteroid cataclysm that doomed the dinosaurs.

In the three and a half years subsequent to her first successful dredge off Portugal, the *Challenger* time and again dredged up *O. lymani*. This poster child of brittle star resilience emerged—in lit-up bits and pieces—from all quarters of the great oceans, from the coast of Virginia to the remote Azores, to New Zealand, to the Sea of Japan. Spawning, bottom-dwelling *O. lymani* sends its offspring toward the surface to be picked up by ocean currents. A young brittle star might settle miles from its progenitor and send its own offspring farther again into new territory. Burrowed in the soft sediment of ocean floors worldwide, brittle stars can achieve densities of 4,000 per square meter. In parts of the Florida Keys, up to half of all marine macrofauna are ophiuroids.

This unstoppable colonizing energy of the brittle star has preserved it through eras of intense ocean stress. During periods of ocean warming or cooling—or sudden dips in oxygen—the brittle star, with its adventurous larvae, has always managed to find safe haven somewhere across the vast, borderless swathes of the ocean floor. And from that redoubt, when conditions stabilized, colonization began again without natural check.

Until now. Wyville Thomson's *O. lymani*, having faced fluctuating temperatures in its long species history, might be expected to survive our current era of ocean warming through its usual tactics. But the ancient brittle star is facing an entirely novel challenge to its hundred-million-year reign. A different rogue offshoot of human industry—microplastics in the marine food chain—poses a threat that no number of self-detonating limbs or expeditionary larvae can reliably defeat. The *Ophiomusium*—gifted with a simple, adaptable design—has resisted all pressures to alter its form or diet since at least the Cretaceous. Natural selection, for instance, has never seen reason to endow the brittle star with eyes. Thus, the most obvious power to discriminate between food particles and plastic in the brittle star's turbid deep-sea surrounds is denied it.

In the early 1970s, British researchers returned to the wild seas off Scotland made famous a century before by Wyville Thomson's pioneering expedition in HMS *Porcupine*. The *O. lymani* that Thomson first identified at Rockall Trough were still there in great numbers, shucking off their arms in the raw air on deck. No thermal energy could possibly reach the brittle star at the depths from which they were recovered, so reproduction must depend on nourishing plankton and fecal food pellets descending from the surface to the ocean floor.

The contents of the brittle star's sack-shaped stomach confirmed its reputation for scrappy resilience. The *O. lymani* hides away in burrows or under rocks until feeding time, then uses its sinuous arms, twisting and coiling, to scurry to an advantageous spot. Employing its signature limbs like an elephant's trunk, the brittle star feeds on any scraps it can find on the ocean floor or floating in the water column above it, from tiny algae to fragments of crab to entire sea worms. Microscopic teeth from any one of its five jaws tears off chunks of its prey (mostly carrion), which it then gulps down whole. Because the supply of food is irregular, both quality and quantity are vital. Plastic, by this measure, is as good as poison to the scavenging, deep-sea brittle star.

For seventy years, shiploads full of human-made plastic have been dumped into the oceans—100 million metric tons annually, on average. Dirty, plastic junk is a visible pollution issue on beaches worldwide, as any disillusioned seaside tourist can attest. Turtles entangled in abandoned fishing nets and sea birds strangled by six-pack holders have become signature images of the ongoing tragedy of marine life in the age of plastic waste. Media attention has focused likewise on the enormous accumulations of trash in open ocean gyres, hundreds of miles wide. Birds and fish have been recovered dying of starvation from feeding on the gyres, their stomachs filled with morsel-sized microplastics they mistake for food.

Then, a further shock. In the mid-2010s, it became apparent that the plastic input to the oceans since the 1950s—an estimated five trillion individual pieces—did not match the observed rubbish accumulating along coastlines and in the great plastic gyres of the Atlantic and Pacific Oceans. A "tremendous loss" of macroplastics—bags, bottles, toiletries, and packaging of all kinds—must be occurring from its disintegration into tiny particles and sinking to the ocean depths.[5]

No less than 92 percent of all ocean plastic, it turns out, is *micro*plastic; for this, the destined landfill is the seafloor and the stomachs of the creatures who roam it. From the moment plastic trash arrives in the ocean—via drain or river or dump truck—natural forces combine to break it down. The wearing power of sun and air, with the constant buffeting of wind and waves, embrittles plastic surfaces, prompting them, like a stressed ophiuroid, to break into ever smaller pieces. The adhesion of organic particles or plankton to the microplastic fragments then increases their weight to a tipping point where they sink through the water column, often indistinguishable in size and shape from the nutrient snow of dead plankton that descends to feed hungry bottom dwellers. Once in the sunless, airless surrounds of the deep sea, the microplastic fragment is protected from further disintegration and may persist in the deep seabed for thousands of years.

With attention focused on the impacts of plastic pollution on picturesque beaches and charismatic marine life, it was not until 2016 that researchers thought to investigate the impacts of sunken microplastics on deep-sea invertebrates. For this, they returned to the iconic waters of the Rockall Trough, first dredged by the *Porcupine* in 1869, and selected, as a research specimen, the redoubtable *O. lymani*. To their surprise and

5 Eriksen et al. (2014) 1.

alarm, the density of microplastics in the water column in Rockall Trough was equal to that along the visibly polluted coasts, threatening doom to the brittle stars. The fragile beauty of *O. lymani* vanishes quickly when raised to the surface. This and its penchant for self-destruction have disqualified ophiuroids from the aquarium trade as well as from commercial fishing. If a billion brittle stars quietly starved on the ocean floor, would anyone really notice?

Death by plastic may be quick—or painfully slow. Ingestion of microplastics can block or perforate a brittle star's stomach, transfer organic pollutants from ocean water to the animal, or leach chemicals into its cell tissue. Symptoms of a plastic diet include ulcerous sores and lesions on the skin. The ophiuroid's signature survival tactic—autotomy—requires energy and thus a steady supply of rich food. Even in the rough seas of Rockall Trough, far from land, the delicate stomachs of Thomson's brittles stars were found lined with a buffet of plastic polymers and microfibres.

O. lymani, dredged by HMS *Challenger* the world over in the early 1870s, entertained Wyville Thomson and his colleagues with its zany brittle star antics. The five-pointed star impressed them, too, with its epic resilience across time and space. But its future is cloudy. Dining on twenty-first-century ocean floors lined with nutrient-free microplastics, will brittle stars lose the vital energy to continue?

In the first days of 1873, the explorers of HMS *Challenger* sailed on an ocean free of plastic, and with minds equally free of any notion of it. But progress was slow out of Lisbon. The ship sounded and dredged all day, making sail only at night on the lightest of breezes. While they waited long hours for the dredge to resurface, the naturalists set out in small boats armed with nets to skim the surface for plankton riches. Highlights were few. Once a sea turtle, fast asleep, bobbed past them but woke just in time to avoid the grasping net.

To their dismay, the newfangled dredge continued to bring up mostly mud until—courtesy of a suggestion forever after credited to the captain—a humble shrimpers' trawl was tried instead. The trawl, on its first try, emerged from the glistening sea full of creatures never seen by professional naturalists—jellyfish and starfish, scuttling crabs, fish with bulging eyes on their backs—as well as a tangled mesh of fern-like soft corals that illuminated the deck with a pale, lilac-tinted light.

Wyville Thomson stood spellbound with the rest. The Gorgonian soft corals, languid and phosphorescent, raised visions of "the wonderful state of things beneath," he wrote in his journal. The ocean floor under *Challenger*'s hull, 500 fathoms deep, "must be animated fields [of wheat] waving gently in the slow tidal current and glowing with a soft diffused light, scintillating and sparkling on the slightest touch, and now and again breaking into long avenues of vivid light indicating the paths of fishes or other wandering denizens of their enchanted region." They brought the corals below deck to the laboratory and prodded them with their fingers, to watch them glow a final time in the darkness. From the unsightly pile of ooze and dead things delivered by the trawl, it was a glimpse of the fantastical world they had been sent to explore.[6]

Among the fish from this spectacular haul, Thomson recognized a slender mora, familiar from Mediterranean seas, as well as the deep-sea grenadier. Their sorry condition—eyes blown out and bladders burst—signified capture at great depth. When the trawl ascended, the ocean's grip slackened, allowing the air in the fish's lungs and tissues to expand fatally fast. All were dead long before being heaved onto *Challenger*'s deck.

The moonlit sky and atmosphere of general relief kept them awake past midnight. They smoked cigars and reveled in the day's catch. The captain had put *Challenger* before the breeze, all sails set, to make up time to Gibraltar, but standing on deck

6 Thomson (1877) 1:119.

in the starry darkness they experienced not the least sensation of motion. The only clue to their onward progress was the appearance of the light at Cape Trafalgar "of glorious memory," twinkling brightly until it too faded. Later, in the soft gloom of sunrise, the coasts of Europe and Africa appeared in a single frame, fronted by black mountains upreared like great stone guards of the ancients. The night clouds vanished ahead of the rosy-fingered dawn. Then, after a few minutes' suspense, the massive white rock of Gibraltar revealed itself to them, "grand and weird."[7]

7 Thomson (1877) 1:126; Wild (1878) 11.

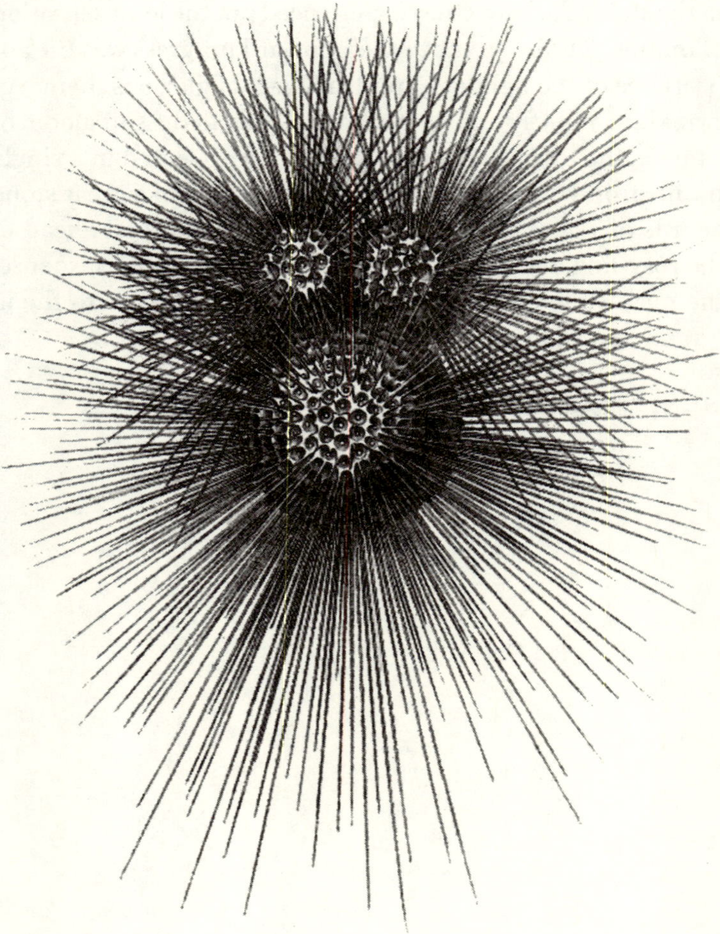

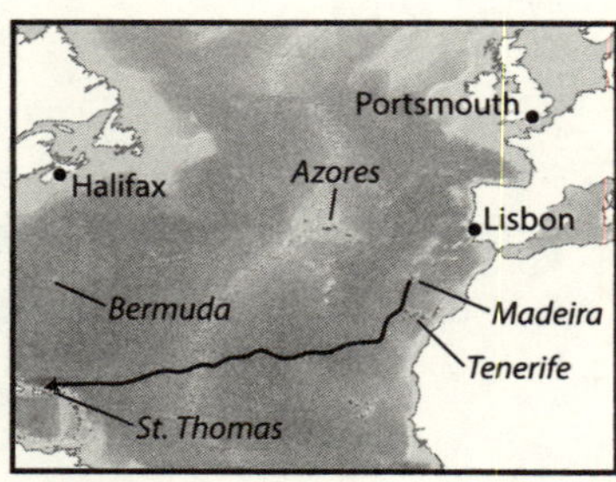

FIGURE 2.1 *Globigerina bulloides*, a plankton with worldwide distribution. *Scientific Results: Zoology*, vol. 9, plate 77 (1884). (Inset) HMS *Challenger* Leg 2. Madeira to St. Thomas, Virgin Islands. February–March 1873.

CHAPTER 2

Skeletons from the Ooze

Tenerife, Canary Islands
28°16′ N, 16°36′ W

Inspiration for the *Challenger* expedition began with an enigmatic "flower" plucked from the depths of the Norwegian Sea in 1868. This sea lily (a class of echinoderm called a crinoid) was not supposed to exist. Its slender, tapering form populated the fossil record of the Cretaceous, but few deep-sea specimens were known to Western naturalists until the Norwegian father-son team, Michael Sars (1805–1869) and George Ossian Sars (1837–1927), set about dredging the cold waters off the icy Lofoten islands.

Hearing news of the discovery, *Challenger*'s future director, Wyville Thomson, took himself from Queen's University in Belfast, where he was chair of natural history, directly to Norway—no light undertaking to inspect a single specimen. If this ancient stalked crinoid lived in numbers on the deep seabed what other living fossils lurked there? And would this modern crinoid—dubbed *Rhizocrinus lofotensis*—exhibit "modification by descent" from its fossil ancestors, proving Darwin right about evolution? Thomson returned convinced that the new frontier for Victorian beach philosophy—its "land of promise"—was the blue wild of the open sea.[1]

Events that followed vaulted Thomson to a career pinnacle in British science and ultimately launched HMS *Challenger*. It happened that Thomson's research collaborator on his

1 Thomson (1873a) 49.

favorite crinoids was William Carpenter (1813–1885), a vice president of the Royal Society. In the spring of 1868, Carpenter lobbied the Admiralty—specifically George Richards, the well-connected hydrographer—at Thomson's behest to commandeer a vessel for deep-sea exploration off the Irish coast, to the Faroe Islands, and south to the Bay of Biscay.

The ensuing missions of HMS *Lightning* and HMS *Porcupine*—led by Thomson, Carpenter, and conchologist Gwyn Jeffreys—made stunning, unanticipated discoveries: first, that the ocean depths were not unmoving and set at a constant temperature, as had been assumed; and second, that the layered sections above functioned like a submarine river system of warm and cold currents. The conspicuous upper ocean Gulf Stream, Carpenter decided, in fact belongs to a global oceanic circulation that funnels cold water from the poles to the tropics and distributes heat across latitudes. These different undersea climatic zones—specifically, their temperature extremes—appeared to govern the distribution of marine animals.

Also, North Atlantic marine life was stunningly diverse. Bottled specimens of colorful echinoderms—starfish, sea urchins, and crinoids—filled the cramped holds of the *Lightning* and the *Porcupine*, providing Thomson with rich material for his book-length report on the missions. *The Depths of the Sea*—a classic of nineteenth-century oceanography—debuted in London bookstores as HMS *Challenger* crossed the Atlantic for the first time in the spring of 1873.

Written in an elegant, accessible style, Thomson's book introduced the wonders of the deep sea to a wide audience and whetted the public's appetite for what other "weird monsters" *Challenger* might bring back from the far reaches of the world. From our vantage point, *The Depths of the Sea* stands as the culmination of a half century of fitful but foundational progress in marine science. The deep sea—barren, cold, invisible—had never before warranted a place in the popular imagination, let alone as a national research priority. Now a

British warship, manned by 240 men, was to spend four years exploring the remote oceans—at the public's expense.[2]

To the *Challenger* sailors, their progress across the monotonous surface of the Atlantic Ocean seemed maddeningly slow. But in the naturalists' workroom on the main deck, the pace was often frantic, particularly late in the evening as the precious natural light faded. It was as if the awesome pressure of the deep sea communicated itself to *Challenger*'s philosophers. Each felt the burden of his appointment to this one-off global odyssey, to bring back a floating museum of deep-sea novelties with which to astonish the world.

Their early soundings showed the shallow continental shelf extending little more than thirty miles from the Portuguese coast before falling away into the deep abyssal plain of the true Atlantic. As a principal objective of their global journey, Thomson's research team sought to raise dinosauric creatures—"living fossils"—from the ocean abyss. The seabed, they reasoned, was a sunless wasteland and must be ancient. The swirling sediment and sunlit garden shallows that made coastal waters a theater of constant change did not hold sway in the deep. Darwin's missing links might lurk in the darkness there. But if the ocean's floor was a permanent and all-weather foundation of the world, its surfaces were transient zones at the mercy of wind and weather. To this extent, HMS *Challenger*, a transient herself, belonged there but was a stranger to the rest, to the Great Beneath. *Challenger*'s mission was to encounter the strangeness of these ocean depths and render them familiar.[3]

2 Thomson (1873a) 49, 157.

3 The Victorian theory of an unchanging deep-sea environment has not stood the test of time. Tectonic disruption ensures the constant, slow-motion renovation of the seafloor along submarine ridges, while recent research shows that even deep-sea fauna have been vulnerable to mass extinction events triggered by climate change.

For *Challenger*'s crew, the case was entirely different. It was an indignity to see the dredging platform—scrubbed to perfection by their own hands before dawn—feet deep in the afternoon with a deposit of freezing ooze like so much sago pudding. It was arguably worth it, however, to watch the well-dressed philosophers wade in, little spades in hand, to shovel the muck from one sieve into another, each finer than the last, until the tiny shells that lined the ocean bottom were separated from the sediment that entombed them. From the distance of a ship's deck, the ocean could seem a blank, unchanging environment. The stunning plankton diversity brought up in *Challenger*'s nets put the lie to that shallow notion.[4]

The life histories of these abundant shells—and of the single-celled plankton they housed—were controversial aboard *Challenger*. Wyville Thomson—a man partial to establishment opinion—believed them lifelong residents of the deep sea. John Murray, however, once *Challenger* was far out from the European coast, discovered live microorganisms in the tow net he used to skim the ocean's teeming surface. One memorable morning, he dipped a teaspoon into the water over the side of his small boat and was amazed to find a living *Globigerina* floating there "in all its glory," protoplasmic body bursting out of its shell.[5]

As *Challenger* crossed lines of latitude, the character of the surface animals changed as did the ocean floor—but not always in concert. In warmer waters, as *Challenger* headed south, Murray found that washing his tow net in a glass vessel left a residue of *Globigerina* and other ubiquitous foraminifera at the bottom. Under the microscope, he observed the fleshly orange-yellow protoplasm of the ever-present *Globigerina bulloides* exude from pores in its beautiful spiral shell.

4 To spare the main deck the worst of the mud, a special platform was installed forward of *Challenger*'s mainmast for the scientists' initial sorting of the dredge's contents.

5 Herdman (1923) 77.

From there, the oozey body spread itself among the forest of spines that protected the whole (or were perhaps a flotation device).

Later, as *Challenger* cruised over the abyssal waters of the mid-Atlantic, Murray continued to collect living *Globigerina* from the surface, while their calcite shells vanished mysteriously from the ooze. The dredge, in these deepest regions, delivered up only an enigmatic red clay, devoid of life. Now Murray and Thomson were equally baffled. Why did the dredge deliver a crate's worth of foraminifera shells onto the deck one day and the next day none at all? Sometimes, they found other wonderful miniature creatures in the clay, a different breed of microorganisms called radiolaria made of silica rather than calcite.

After nearly four years tracking the appearance and disappearance of *Globigerina* ooze across the world's oceans, it was *Challenger*'s destiny to settle both the debate over the foraminifera's habitat and the mystery of its disappearance. Other Earth secrets hidden in the foram shells have since been revealed, including the changing temperature of the seas over millions

FIGURE 2.2 *Globigerina* remains blanket the seafloor with its signature "ooze." Here seen under a Victorian-era microscope. *Scientific Results: Deep Sea Deposits*, plate 11 (1891).

of years and the origin of the Asian monsoon. Not everything is an open book to us, however. Tiny plankton, en masse, represent the largest single biomass in the oceans. But whether the ubiquitous calcite-shelled *Globigerina* and its fellow forams—keystones of protistan life since the Cambrian—will survive the acidifying oceans of the Anthropocene is an open question.

At Madeira, *Challenger*'s officers strolled among English gardens full of imported roses, geraniums, and nostalgic oleanders. Meanwhile, on the dock, resupply proved a treacherous business. The men offloaded sacks of coal from the boats in a heavy surf. A brisk trade wind out of the northeast then carried *Challenger* down the African coast on the maritime highway to Tenerife—their last port of call before the West Indies, 2,500 miles to the west. Where Madeira had offered green woodlands and scents of Europe, Tenerife, dry and rugged, seemed like a little chunk of Africa drifted out to sea. At first light, the land they knew existed was obscured by rain. Then the snowy peaks appeared above the clouds, as if floating there.

They entered the harbor at Santa Cruz at dawn, with steep rocks close, and the mountain peak invisible behind a curtain of clouds. The terraced slopes of the famous volcano appeared to them like a blanket of white blossoms. Alexander von Humboldt—first of the European naturalists to make this pilgrimage to the Canary Islands—had met with lush banana groves and strawberry vines down to the shore. But these had been ripped out and replaced by fields of an imported cactus wreathed in white cloth.

The *Challenger* philosophers visited these plantations on their first day, where they witnessed a winter crop of insects (specifically, cochineals) introduced by the thousands to feed and breed on the lobes of the cactus plant. The fattened cochineals, pulverized into their signature crimson dye and stacked into barrels on the waterfront, brought high prices on the European textile market.

FIGURE 2.3 An early nineteenth-century view of Santa Cruz, Tenerife, with the Mount Teide volcano in the background. Drawn by W. Alexander and engraved by T. Medland from John Barrow, *A Voyage to Cochinchina in the Years 1792 and 1793*, plate 3 (T. Cadell & W. Davies, 1806).

The French consul at Madeira—an amateur botanist—pointed proudly to an image of Alexander von Humboldt hung on the wall. It was the famous naturalist's well-known library portrait, surrounded by specimens, books, and maps. In the foreground of the painting, a reproduction of the globe whose oceans the *Challenger* proposed to sail for the next four years sat prominently on Humboldt's desk. Medals bestowed by royalty and the scientific societies of Europe lay casually strewn about. These were the rewards of fame to which Murray, Moseley, and the rest of the *Challenger* naturalists aspired. The consul, meanwhile, approved their plan to follow Humboldt's footsteps up the Tenerife volcano but warned no guide would accompany them beyond the snow line in winter.

They rode in carriages through the white fields to Orotava where descendants of the Spanish conquistadors kept their luxurious villas. After lunch, they visited a little museum dedicated to relics of the indigenous Guanches, exterminated centuries before in a one-sided struggle with European invaders. The native communities not massacred or decimated by disease had been sent across the ocean into slavery. Not a single living representative of the race remained on the island.

From Orotava, the iconic peak rose like a giant shard into the blue sky. A dense line of clouds girdled the slopes halfway up, transported by trade winds and fueled by an infinite supply of vapor from the surrounding ocean. Above the clouds, the sun danced brightly on the white peak. Beneath, a green valley wound down to the island's rim, edged with snow-white surf. They ascended the volcano, mindful of Humboldt's principles of plant geography. To climb above sea level into colder elevations was no different than a ship crossing lines of latitude on the ocean's horizontal plane. As the temperature changed, so did the nature of the plants the climate supported. They agreed their ascent of the Tenerife volcano was perfect preparation for a crew of oceangoing philosophers.

Humboldt had promised that, in the course of a few hours' climb, they would travel from the tropics to the poles. But the reality was more complicated. Beyond the last line of cottages, with their vines and fruit trees, they entered a wilderness of heather that ended only with their ascent into the cloud. Emerging from the mist, they discovered a moonscape of rocks and desert. Beyond it, only ugly scrub and a tiny tenacious violet (*V. cheiranthifolia*) continued the struggle for existence. Up and up, toward the craterous peak, all vegetation disappeared. The heat burned their faces, while the snow froze their feet into their boots.

The moon came up early, bright enough to read by. Looking out over the cloud was like scanning the ocean from *Challenger*'s masthead. At intervals, the wind over their shoulders drove a cleft in the ocean of fog to reveal the actual sea a mile below. They glimpsed an inverted heaven framed by mist, stunning blue and flecked with white. The streets of Santa

Clara far below had been hot enough for camels. Here, on a volcanic ridge above the town, they lit fires in the desolate brush to keep from freezing.

For Murray and Moseley, it was a lesson well learned. In the oceans—as on a mountain—climate was vertical. The *Challenger*, with its dredge and trawl, was to sound distances beneath the sea as great and greater as they had climbed above it. Would clear, Humboldtian zones of life emerge from the deep sea according to temperature and depth, in the same way organic life had organized itself on this mountain? Would they find divisions of life as abrupt as the specimens they gathered progressively from the slopes of Tenerife—first palm fruits, then pine and heather, and then only moss and lichen? And finally, beyond an unknown threshold of cold in the dark waters, would they discover life extinguished altogether, as here on the bare, moonlit rock below the volcano's summit?

Whatever they might find or not find, Murray and Moseley took comfort in one thing. On Tenerife, they had seen enough of the ravages of human-driven extinction. Plants uprooted, trees cleared. New crops and animals shipped in from abroad. An entire race of people brutally extinguished. The ancient creatures of the deep sea, however, would still be waiting centuries hence for future explorers—so they thought. Not gifted with foresight, the Victorian philosophers confidently assumed that whatever revolutions an ambitious humanity could wreak on the Earth's surface lands, the oceans stood inviolably vast and beyond the reach of human meddling.

They returned to find *Challenger*'s decks filled with flowers and fruits from the island. Bright green canaries, set loose in the rigging, flew about their heads. They rose early in the morning to see the dredge put over the side. Once it was judged to have touched bottom and filled, the *Challenger* hove to. Then the drawn-out screeching of the pulleys commenced and continued all day, as it would every second day at sea on average for the next three and a half years. When at last the bulky bag of the dredge appeared, it was obvious something was amiss. It had gotten tangled in the rope while paying out and was empty. So Murray occupied himself through the night with the gritty planktonic remains from the tow nets dropped over the side.

For a thousand miles of Atlantic Ocean, before coming to anchor at St. Thomas, a great white shark menaced HMS *Challenger*. The animal was intelligent and would never bite at the hooks baited with pork along the ship's side. In a change of tactics, they floated a parcel of meat astern on a raft. But the cunning monster waited for the ship to pull away and then devoured the float in a frenzied splash, showing off its great jaws. The officers fired their rifles at it from the upper deck. The shark responded by taking refuge beneath the ship, out of sight. The sailors on *Challenger* nurtured a deep, inherited hatred for sharks. If it *was* somehow caught, the men promised each other to eat it alive, sliced up in four-pound morsels. But more days passed with *Carcharodon carcharias* as *Challenger*'s stealth companion. Finally, the men agreed it would never leave them alone until someone onboard was dead and thrown overboard to satisfy its malice.

The shark was not a lonely hunter. Pilot fish swam alongside waiting for scraps, while the tenacious suckerfish remora clung to its belly. From the deck, Moseley studied the weird association—three fish, utterly different in character, gliding about the ship "like a single compound organism." They seemed to mistake *Challenger* herself for a giant shark. Day and night, the pilot fish darted just ahead of the ship's bow, which they presumably figured for a snout. Only after several days did disillusionment set in and the hungry fish seek their fortunes elsewhere.[6]

Meanwhile the remora, bane of all superstitious sailors, clung to *Challenger*'s hull as if it were a leviathan's belly. They

6 Moseley (1892) 8.

caught a foot-long specimen and laid it on a table. It could not be pried off. Legend had it that the sucking powers of a single remora could reduce a thousand-ton ship to paralysis. No progress, even in a gale, until the bad-luck-bearing parasite was removed.

They had success catching several smaller sharks and hung them suspended above the deck. The ship's two pet spaniels growled and had to be held back from the still-snapping jaws. They ate these fierce victims for breakfast: tasty but dry. Overall, the *Challenger* scientists were disappointed in the lack of species variety among the Atlantic sharks. Dredging the ocean bottom had brought up piles of remarkable predator teeth in the ooze, evidence of a richer shark kingdom that eluded them. Or did these outsized incisors belong to giant monsters of oceans past?

The company of the shark during their Atlantic crossing distracted at least from the relentless tedium of dredging. "Drudging," the men called it. The days after departing Gibraltar had been sunny and delightful. They had picked up the trade winds on cue and might have made a very brisk leg but for stopping, every other day, to churn up the ocean bottom and deliver tubs of cold mud to the assembled philosophers. The small engine on deck bellowed and spluttered ten hours at a stretch. Often the dredge emerged in darkness. From the naval point of view, it was another day lost idling the ship while the precious trades whistled tantalizingly through bare masts. This unnatural proceeding, in defiance of all seagoing instinct, so unsettled the men that finally Captain Nares made a personal request of Director Thomson to address the ship's company on the subject of the expedition's purpose.

So they gathered in the breezy evening air on deck, vaguely ill-disposed to the expedition but not to the professor himself, who seemed a harmless, if unseamanlike, old gentleman. The oceans, he told them, covered nearly three-quarters of the globe but were like a sealed book to the human race (this was not news). Nothing could exist below 400 fathoms, it had long been believed, so the deep sea was of little consideration until the great nations of the world decided to build a long-distance telegraph as a means of communicating with each other and released funds to research the practicality of submarine cables. The deep sea, at last, was interesting to politicians.

Now, only three months into their voyage—the professor announced—*Challenger* had already achieved the deepest sounding in history, over 3,000 fathoms (5.5 kilometers). They had discovered animal life in abundance at those depths, which put the lie to Forbes's azoic theory. Even the apparently lifeless mud far below had yielded up to John Murray's microscope the shells of recently living creatures, in the infinitely slow process of forming a chalk bed akin to the white cliffs of Dover. His *Challenger* colleagues could already confirm, moreover, that the very ocean beneath them, even at moderate depths, was "teeming with animals of a sort unknown to man: animals nearly transparent, but which have eyes, and lungs, and hearts the same as we have."

Thomson pointed to a giant blackboard image of a sea polyp drawn by Jean Jacques Wild, the ship's artist—it was half plant, half animal, and emitted a luminous green light from its body. How the tiny creature resisted the enormous pressures of the ocean at depth was yet a mystery, Thomson explained. If a man were submerged on the ocean bed three miles below, where they had sounded that day, "he would have a weight pressing on him equal to the weight of the *Challenger*" herself. This was a vivid illustration the men had reason to recall weeks later when they assembled on the same deck to watch the body of one of their own slide through a porthole into the deepest abyss of ocean known to man. For now, however, they only grinned and applauded the first and last scientific lecture given aboard HMS *Challenger*.[7]

7 Matkin (1992) 60, 58–59.

German Darwinist Ernst Haeckel was the first to identify single-celled Protista as a separate kingdom of marine life independent of animals and plants. As a vacillating young man on an Italian tour in the early 1860s, he briefly toyed with abandoning science for a career as an artist. Then a German translation of *The Origin of Species* fell into his hands. In it, he found the basis for a pure, materialist worldview purged of religious obfuscation.

Looking out across the straits of Messina, teeming with surface life, Haeckel saw in an instant how the "sympathetic selection of colour" had made its residents transparent as glass or blue like the water itself to protect them from predators beneath. This mark of adaptation, moreover, was consistent across all classes of open-sea animals, from mollusks, to snails, to worms, to the ubiquitous jellies. He intuited, too, the deep connection between climate change and evolution: how "change of climate must have affected the geographical distribution of organisms and the origin of numerous new species." Natural selection, as an explanatory framework for the history of life, was for Haeckel all powerful.[8]

The Victorians' hunt for the origins of life was tantamount to obsession. Haeckel discovered it in the scummy surface of the sea, home to an "exquisite variety" of single-celled protists. Among these, several classes had the ability to secrete adjunct skeletons in wonderful crystalline shapes, visible under the most advanced microscopes of the day. The sight of them, he wrote to his father, left him "lost for words with joy and delight." Haeckel was gifted with a genius for multitasking. He could study a complex, single-celled animal through a microscope while sketching it with dazzling exactitude with his free hand. His famous illustrations captured the previously undiscovered microorganic world with a flair and precision that astonished Darwin himself. *Natura in minima maxima.*[9]

8 Haeckel (1887a) 1:264, 1:364.

9 Quoted in Gregorio (2005) 66; "Nature is greatest in the smallest things."

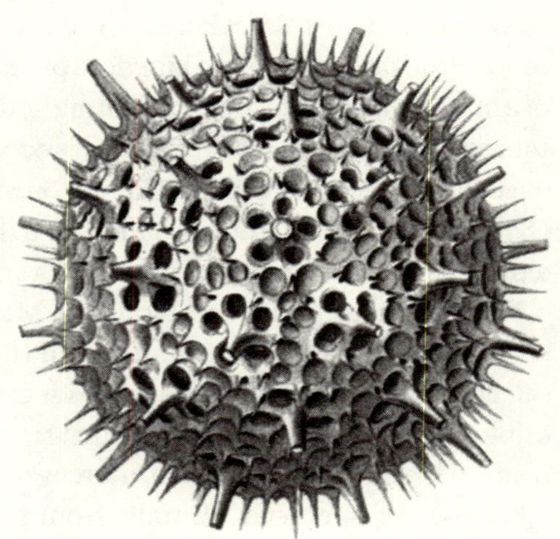

FIGURE 2.4 *Haeckeliana darwinia*, from Ernst Haeckel's famous suite of radiolarian drawings. *Scientific Results: Zoology*, vol. 18, plate 114 (1891).

The wonder of the radiolaria for Haeckel was that the full rich business of life could be carried out by an animal one-cell in dimension; that animal family, moreover, was comprised of infinite variations on a simple, colorless sphere. A plant sustained itself by synthesizing chemical compounds necessary for its sustenance, but the radiolarian, like any other animal, depended on its environment for nutrition, absorbing oxygen and exuding carbon in turn.

Long, thin filaments, called pseudopodia, issue through pores in the gelatinous mantle that surround the radiolarian body. These thread-like appendages are the animal's contact points with the outer world. The pseudopodia breathe, ingest food, sense threats, and register like and dislike. The tentacular feet likewise enable propulsion, of a sort. Haeckel observed these "slow, oscillating, sinuous motions" of living radiolaria captured in a jar.[10]

With a body density just greater than water, the radiolarian circulate upwards and downwards through the water column

10 Haeckel (1887b) cxli.

in a kind of managed drift. This lifesaving friction (to prevent sinking) is enhanced by a spectacular armory of spikes organized about the spherical body. These bristles, which more than double the creature's physical area, are designed to ward off predators, Haeckel guessed, but also to maintain bodily equilibrium amid the relentless aquatic turmoil of eddies, swells, and currents. John Murray witnessed this spontaneous protistan ballet himself in his specimen jars aboard *Challenger*. When the radiolarian body later bursts in a sacrificial orgy of zoospore offspring, its skeletal armature—all that remains—possesses built-in stabilizers for its long death spiral to the ocean floor miles below.

Through the 1860s and 1870s, Haeckel was Charles Darwin's most active champion in Europe in the debate over evolution, with the radiolaria his exhibit A. What else but the modification of species over time, he argued, could explain the architectural variety of radiolarian skeletons—one like a stunning ball gown, another like armor mesh, another like honeycomb? Only the relentless, subtle hand of natural selection, over geological time, in the ever-shifting conditions of the sea's surface could have produced the "boundless firmament of stars" under his microscope.[11]

Answers to the mysteries of radiolarian life lay in the mysteries of radiolarian form—the most extravagant parade of shapes in nature. In the delicate geometries of the protistan skeletons, Haeckel perceived a "fundamental form" of nature common to both organic and inorganic realms. What else could explain the durability of these tiny creatures whose living masses dominated the oceans, and whose fossilized remains formed mountains and entire islands? The sheer wealth of protist species also offered proof of a basic Darwinian principle: the most successful creature types would spawn the greatest variety of forms.

11 Haeckel (1887b) i.

Ernst Haeckel did not sail aboard HMS *Challenger*, but he received crates full of its bounty in the mail on the ship's return, to be studied and written up for publication. A detailed analysis of more than 4,000 radiolarian species required ten years. Haeckel's long, Darwinist essays on radiolaria, deep-sea sponges, and gelatinous hydroids (the siphonophores) are among the highlights of *Challenger*'s fifty volumes of scientific reports, each the size of the Guttenberg Bible and recognized as founding scriptures of modern oceanography.

While the professor, Moseley, and Willemoes-Suhm spent long evenings dissecting the larger, charismatic animals from the dredge, Murray continued to concentrate his energies on the microscopic creatures of the mud itself. What swam? What sank? What of the graveyard of shells on the ocean floor, yards deep? Did these animalcule legions live and die unseen in the dark, parental mud, as consensus had it, or did they sport miles above at the sunlit surface and descend only at death in a thick, microbial mist, bodies dissolving en route, their vacant shells adding layer on layer to the ever-deepening strata of deep-sea ooze?

Leaning over the side, Murray watched the ship's prow cleave great, smeary chains of protists on the surface. He had the boat lowered to bring back more jars full of tiny spinning spheres to the laboratory. Under the microscope, he thought he detected dozens of new radiolarian species—an alien patisserie of cones, frills, and lattice meshes to be stained with carmine and oh-so-carefully mounted in balsam for future study.[12]

12 Ernst Haeckel's tome on the *Challenger* radiolaria was published, after ten years' labor, in 1887. Henry Brady's report on the foraminifera—half the length but still weighing in at 750 pages—appeared in 1884. Both are seminal documents of their respective fields. Likewise, John Murray's map of global seafloor sediments and their distribution was not superseded for almost a century—a remarkably accurate first edition of its genre.

The sun safely set, a million creatures played over the surface of the sea, lighting the *Challenger*'s wake. Some nights, "there was light enough to read the smallest print with ease," and the philosophers composed their daily journal entries on deck. The same creatures then descend by day, Murray presumed, to the sanctuary of mid-water gloom, away from cruising sea birds and surface predators. If he could only catch them in the act—this living mass on its daily vertical commute, distributing oxygen, nutrition, and life itself to all sectors of the ocean, even down to the very abyss.[13]

For the long weeks of their Atlantic crossing, during their nightly gathering in the mess to smoke and philosophize, Murray debated Thomson on the origins of the pasty shell remains on the ocean floor. The ubiquitous, single-celled protists—the foraminifera, radiolaria, and diatoms in their millions—lived free-floating at the surface, he was now convinced, before sinking slowly to the bottom at their death. The ocean water column played host to this epic dying fall—a constant descent of tiny shells, like snowdrift, to the bottom.

But still the professor resisted. So much easier, Thomson argued, to imagine the plankton living where they died in the abyssal darkness than the survival of their shells through miles of shifting currents and all the sudden changes in temperature and pressure. Thomson was committed to his vision of an ancient, unchanging ocean floor, like a Cretaceous fossil bed.

The radiolaria impressed both Thomson and Murray with their sheer evolutionary difference from humans. These tiny protists lived as they themselves lived, performing the same functions of life and bound by the same physical laws. And yet the radiolaria and the *Challenger* naturalists who studied them inhabited opposite extremes of the planetary spectrum of creation—brought together only by historical happenstance in the dim light of the *Challenger* workroom. Even in their

13 Campbell (1877) 39.

captive jars, the radiolaria appeared wild and inhuman beyond conception. Between a planktonic cell and the man who peered at it via the length of a microscope opened a gulf of existence deeper than any ocean.

Time—the great agent of radiolarian variety—also ensures their universal presence in the world's oceans. For hours at a stretch, in an open boat with the sun beating down, Murray followed the planktonic streams as they drifted miles with the Atlantic current—a grand, passive migration. Back on deck at night, he saw millions of them still, lit up in a sinuous, shining layer across the ocean's surface like a winding road into futurity. Everywhere on the *Challenger*'s world-girdling voyage, whether he cut into the belly of a crab or worm, fish or octopus, he found radiolaria bobbing there, staple diet for ocean life. This meant the radiolarians were ubiquitous not only across the ocean's surface but vertically in the water column, to be gorged upon by marine animals of all neighborhoods and appetites.

The very month *Challenger* completed her first Atlantic transit, furling her sails in the calm, blue waters of the Caribbean Sea, Ernst Haeckel's compatriot Christian Ehrenberg published his pioneering study of radiolaria in the fossil record—what he called "micro-geology." Ehrenberg had investigated the unique radiolarian cliffs of Barbados, beyond the horizon to the south. The island was a coral patchwork of ancient ocean beds uplifted from depths below the threshold where carbonate shells dissolved. The inescapable conclusion? The freezing ooze that *Challenger* had dredged from miles beneath the surface of the ocean would, in some future age, be transformed into a beach or forest floor, baking beneath a tropical sun just like Barbados.

How this epic reversal of marine and terrestrial worlds occurred, as it often must have done, Murray could not imagine. Ooze signified life; red clay meant death—or the absence of life. The great plankton colonies at play on the sea's surface flirted with the deep—floating and sinking by turns—before their downward, mortal transit. As they descended, some were eaten by the sea-dwellers lurking beneath the surface, while survivors reached their limits of acidic tolerance and dissolved (a critical threshold now known as the calcite compensation depth). For thousands of miles across the open Atlantic, this accounted for the entire protistan multitude where the dredge scooped only lifeless red clay. Then, as they drew nearer to their Caribbean island harbor, the shell-studded ooze returned as if to remind them of the existence of beaches and men and all their gaudy, tragic concerns.

Contrary to their expectations, the ocean abyss had not proved to be an uninterrupted sedimentary plain. In early March, their soundings—combined with the reappearance of *Globigerina* ooze—indicated they were passing over a submarine mountain range. Over the subsequent three years, they would map the trans-hemispheric course of the mighty mid-Atlantic Ridge from the North Atlantic to Antarctica. Tectonic theory would require another hundred years to emerge, but *Challenger*'s Atlantic soundings were the first to signal the reality of seafloor spreading—a deep-sea geology continually rebuilding and refashioning the continents of the Earth.

Tiny as forams are, their long-term skeletal presence in ocean floor sediments makes for a treasure trove of environmental data, specifically related to climate change. Beginning in the 1950s, researchers determined that the changing distribution of foram fossil species buried in the ocean floor corresponds to fluctuating ocean temperatures over millions of years. Meanwhile, the fragile shells themselves, built from minerals absorbed from the water column, record the state of ocean chemistry at the time of their creation. For example, a preponderance of heavy oxygen isotopes signifies a cold weather regime when lighter water molecules are locked up in mile-high ice sheets.

Then, in the 1990s, another breakthrough. Magnesium levels in foram shells function like an oceanic thermometer, allowing for more precise, long-term historical reconstructions of temperature change. Results include comprehensive proxy datasets for sea-surface temperature change in the Atlantic, Pacific, and Southern oceans over epochs of geological time.

By extrapolation, fluctuating populations of the king of forams, *Globigerina bulloides*, mark the historical rhythm of the world's major weather systems—including its most spectacular, the South Asian monsoon. Floating on the ocean surface, where John Murray found them in scummy abundance, *G. bulloides* thrives in productive ocean regions with nutrient-rich, upwelling currents. These currents correlate directly to the speed and direction of surface winds. A spike in *G. bulloides* drawn from sediment cores in the Arabian Sea has pinpointed the origins of the Asian monsoon about ten million years ago. Modern marine scientists, on foundations laid by Murray aboard HMS *Challenger*, have constructed a historical climate database of protistan complexity and detail.

But the *Challenger* paradox applies here. Deep-sea foraminifera—*Challenger*'s gift to science—have also been a boon to the petroleum industry. Their species distribution—a code for temperature—functions like a floor map of the Carboniferous, leading prospectors to precious deep-sea reservoirs of oil and gas. In a deep historical irony, the so-called great acceleration of human fossil fuel consumption beginning after World War II corresponds precisely with forams' emergence as the great bioindicator of the world's oceans and climate.

There's a final twist in the tale. In 2018, curators at the Natural History Museum in London dusted off John Murray's foraminifera shells from *Challenger* plankton tows and set them alongside their twenty-first-century counterparts exposed to elevated ocean acidity. The results were alarming. The foram shells of the Anthropocene were significantly thinner than Murray's preindustrial baseline specimens. Corroborating research shows that the foraminifera species now accumulating on the seabed are much thinner-shelled than those buried beneath them. The ocean floor that HMS *Challenger* cruised over 150 years ago are not our ocean floors.

Rising acidity in the seas—a direct product of runaway carbon emissions—threatens the viability not only of worldwide foraminifera populations but all calcite-shell-producing organisms, including mollusks, corals and crustaceans. The near-daily spectacle of Murray and his colleagues wallowing in *Globigerina* ooze, sifting for tiny shells, amused the foremast hands of HMS *Challenger* back in the last days of sail. Even then, Murray understood the importance of single-celled marine life to ocean history from its ubiquity in sedimentary rocks from the Italian Alps to the West Indies. Hence the diligence with which he documented the alien variety of these open ocean plankton, while keeping close account also of the delicate seawater chemistry that, altered even slightly, will surely eliminate them from the fossil record of the future.

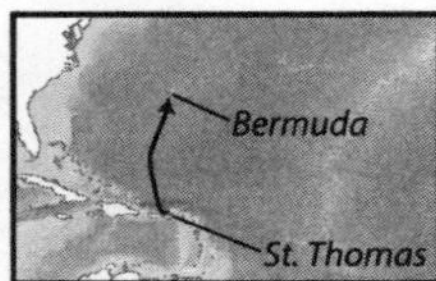

FIGURE 3.1 Once ubiquitous, Elkhorn coral (*Acropora palmata*) is now critically endangered. *Memoirs of the Museum of Comparative Zoology at Harvard College*, vol. 7, no. 1. plate 18 (1880). (Inset) HMS *Challenger* Leg 3. Virgin Islands to Bermuda. March–April 1873.

CHAPTER 3

The Fire Reefs

St. Thomas, Virgin Islands
18°20′ N, 64°55′ W

The annual renewal of coral reefs in the Caribbean begins with a magical synchrony of nature. In early August 1872, while the soon-to-sail HMS *Challenger* rocked at her berth half a world away, the tropical west Atlantic stirred with new life. On a succession of summer nights following the full moon, at a mysteriously agreed-upon hour, miles-long legions of coral polyps, like tiny curled fingers, suddenly jettisoned their eggs into the dark water. The effect, for the few lucky persons ever to witness the spectacle, is like pixie dust sprayed from a million nozzles—an underwater snowstorm.

From the Florida Keys to the Antilles off South America, summer clouds of gametes float to the surface to begin their ritual fertilization. Currents tug them from their mother reef toward open water, while eggs and sperm from different coral colonies mix to promote genetic diversity (and hence resilience). Nature's stern laws mean many never make it—gulped up by hungry predators in the water column or simply drift away. But by the time *Challenger* set sail from Sheerness that December, millions of surviving coral embryos had sunk again to their natural depth, settled, and begun their change of state to a start-up coral polyp complete with mouth and tiny tentacles.

In early spring, when *Challenger* emerged from the Atlantic swell into the sheltered blue waters of the Caribbean, the new corals had entered on a second mass spawning—asexual this time. Fueled by symbiotic algae housed within their cells, the junior polyps calcified into their more familiar skeletal form and began budding copies of themselves, new branches inching upward to the life-giving light. Beneath them along the sandy ocean floor lay the sunken rubble of their ancestors, picked over by schools of multicolored reef fish. From their first hours, the new limestone beds were locked in a race against natural aquatic forces that would eat or erode them. So it had been for the corals of the Caribbean for half a million years, through ice ages and the rising and falling of the sea, when *Challenger* dropped anchor off the island of St. Thomas on a stormy night in March 1873.

A century and a half later, much of the spectacular, real-world aquarium that *Challenger* encountered in the Caribbean Sea is gone, with what's left struggling to survive.

In the weeks before their arrival in the West Indies the *Challenger*'s people sweltered through the equatorial heat. During the day, the naturalists, dressed in their customary suits, strolled beneath a giant awning spread for shade. In the evening, the captain permitted smoking on the upper deck to while away the sleepless hours. From there they could observe the bluejackets at their favorite pastime: fishing over the side, their legs dangling above the coursing water.

A tuna was caught, as well as smaller fish in abundance. The Southern Cross, newly arrived in the night sky, traced its child-like impression on a backdrop of infinite space. Accompanied by a bright moon, it seemed to bless their familiar motions, watch on watch. They had been at sea almost a month. Flowers and fruits had disappeared from the officers' mess, though the fresh poultry supply from Cape Verde was not yet exhausted. The trade wind had abated, but lately the westward current was drawing them into the crystal waters of the Gulf Stream where the nighttime temperature of air and sea drew closer to even—a torpid 74°F (23.3°C).

In this flawless blue ocean realm, entertainment could be had by simply peering over the ship's side at the darting schools of dolphins and flying fish, and at the stinging jellies and the siphonophore *Physalia* scooting by. The sailors called these Portuguese man o' war for their bladders set aloft like sails. One night a flying fish launched itself aboard, attracted by a lantern. Two dolphins were caught and eaten. Even more diverting was a morning sailing match with a small Spanish brigantine, which drew the entire ship's company on deck. When the vessel drew within hailing distance, the Challengers observed a deck full of crates of onions and that the Spanish traders had brought their wives with them. No fewer than six women were counted, sparingly dressed in the heat.

One day, via the perfect transparency of the Gulf Stream, the seafloor itself rose into view. They lifted their gaze, and land appeared at last on the horizon—a blue, misty mass. With their approach to shore, the skies began to fill. Frigate birds soared overhead, tail feathers spread wide. Pelicans dashed down from the sky at unseen prey on the water's rippling surface. When they paused to dredge, they found the mud from the ocean bottom had mysteriously changed character. The grey bed of plankton shells from the open ocean—Murray's *Globigerina* ooze—had given way to rocky residua of the land itself. The millennial work of waves and tides on the shore had drawn terrigenous grains into the ocean's maw.

Challenger's offshore haul of March 15, 1873, was full of eye-catching wonders: corals, sponges, starfish, crabs, and brightly striped fish that none of them had ever seen or read about. They were now among the fabled Caribbean reefs—rainforests of the sea, hosts of the greatest color and variety of life in all nature. Something was missing, however. The rivers of the island would—in ordinary circumstances—have collected sediment from inland, even whole plants and trees, for delivery to the coastal waters. But when Henry Moseley looked from the white beaches of St. Thomas to the contents of the dredge splattered on deck, he saw no bits of trees or roots—no floating logs one would expect in the vicinity of a tropical island. The ooze of the open ocean might be ancient—the professor dared to call it "Cretaceous"—but this Caribbean mud was distinctly modern.

Two short centuries before, the forests of St. Thomas had been cut down to create sugar plantations and the island populated with enslaved Africans to harvest them. This new world agriculture had altered the biology of the island beyond recognition. Then, mere decades previous, with emancipation, the British planters abruptly left, abandoning the sugar fields to a wild bush that appeared, from *Challenger*'s deck, to be the island's native vegetation. It was in truth an invader, like the pigs, goats, and guinea fowl brought from Europe, sprung from captivity by the chaos of perennial hurricanes and now turned feral across the island.

Instead of natural forest residue, the waters off St. Thomas were littered mostly with human detritus: ships and floating crafts of all kinds, wrecked by hurricanes or simply left behind when the plantations collapsed. Descendants of the newly emancipated laborers had improvised living quarters from the broken vessels of their overseers. Moseley could make out the rickety lean-tos lining the scrubby heights above the shore. Closer inspection had to wait, however. The night of her planned arrival, *Challenger* was given a taste of the storm-prone Caribbean climate. Gale winds and heavy rain forced the ship to furl her sails and lay to. When the weather cleared next morning, they found themselves several miles to leeward of their destination and forced, ignominiously, to steam into harbor.

Nestled in a cleft at the foot of two mountains lay the island capital, Charlotte Amalie, with its whitewashed cottages and red-tiled roofs. For more than three centuries, Europeans had traded Caribbean islands among themselves. The Danes now governed St. Thomas, but on the streets English was the lingua franca. Women at the market square wore brightly colored scarves on their heads, recalling Africa.

To curb the summer ravages of yellow fever, the governor had ordered a portion of the harbor reef carved out to promote tidal reflux of the polluted bay. Moseley took a ship's boat through this human-made channel and pulled alongside one of the little wooden jetties dotting the shore. The locals watched him with casual attention, seeming amused. All that remained of the island's plantation crop were little sticks of sugar cane everywhere underfoot, which the dock laborers sucked on while shovelling coal onto *Challenger*'s rear deck—fuel for her continuing explorations across this part-human, part-wild marine world.

Strolling along the beach, Moseley found lines of dried-out seaweed and corals bleached white by the intense sun (only a merciful trade wind from offshore made the midday heat bearable on St. Thomas). On the wet sand by the water's edge, he came across black, spiky sea urchins in great numbers—*Diadema antillarum*. He knelt down to collect specimens, taking care to avoid the bristling spines. But they stung him anyway, a nasty cut on the finger that drew blood.

D. antillarum served as the principal consumer of West Indian corals for at least a hundred thousand years, reliably turning skeletons to sand. But little more than a century after Moseley's visit—Christmas 1983—the same beach at St. Thomas was filled with dead and dying sea urchins. An unknown pathogen borne along by ocean currents was devastating *D. antillarum* all across the Caribbean—the worst epidemic yet recorded for a marine invertebrate.

Sea urchins like *D. antillarum* have a voracious appetite for seagrass and large algae, clearing floor space around the reef for newborn coral polyps to settle and grow. But with their urchin nemesis gone, macroalgae have spread unchallenged through the Caribbean reef system, crowding out vital reef-building corals. Nor has *D. antillarum* recovered; a further mass mortality event struck in 2022. The coral spawn of now four decades past, sinking by their millions to the seafloor, have found nowhere safe to land.

Challenger's regular trawl could not be risked on the razor-edged reef with its infinite traps and tangles. So Moseley spent long days in the small steamboat called the pinnace, equipped with a one-person dredge. Armed with his water glass—a rectangular box with a clear panel at the bottom—he leaned over the side for his first glimpse of the teeming coral garden that began only meters from the beach.

Ivory tree corals, maze corals, rose corals, saucer and cactus corals—in weird floral shapes—crowded among brain corals patterned with the intricacy of a porcelain vase. A little deeper and the sweeping *Gorgonia* rose into view, undulating with each wave like giant fans in the breeze, before the reef sloped away into the gloomy deep. Against this vivid backdrop, the reef-loving fish amazed him with colors not found in terrestrial nature: bright yellows, pinks, and purples, arranged in spots, stripes and other fancy geometries, never still, darting and shimmering—an orgy for land-bound eyes.

He stripped down and dived in, to hover as best he could about the reef in its own element. He kept clear of the notorious *Millepora*—the branching fire corals—with their stinging polyps, but on returning to the boat he felt their effect in his red, raw eyes. When it came to specimens of these awkward fire corals and others inaccessible to the dredge, he hired locals from the docks to fetch them.

One memorable day, Moseley rowed out with a few of the officers to a small, reef-ringed atoll in the channel. They sat among the dead, white coral and shells that had washed up on the sand, and watched the crabs with stalk-like eyes scuttle among the rocks. Broiled by the sun, the Challengers sought relief in the shallows among the living "elkhorn" coral—*Acropora palmata*. They appreciated how its great broad leaves offered shelter to a school of sky blue parrotfish, the very type of shade they lacked.

Everywhere off the beaches of St. Thomas in his little boat, Moseley met with more wide-leaved *A. palmata* and its sister coral, the "staghorn" *Acropora cervicornis*. Eyes glued to his water glass, he studied them through the hot afternoons. The luxuriance of these reef-building corals seemed to him without limit. At the bottom of the sloping reef spread great bramble thickets of staghorn coral. Above these, winding miles along the shore, the antler-like *A. palmata*. Everything that lived on the reef seemed to depend on these massive *Acropora*, from the more delicate corals growing in their shade to the fish feeding on their skeletal branches. If reefs were forests of the sea, these were its ancient oaks, the primeval floor and canopy. Nip, nip, and a brightly striped reef fish flickered out of sight. The little puff of coral dust left in its wake showed how the tumescent *Acropora* were supporting, in real time, the living forces of their own erosion.

Today, fewer than 10 percent of the *Acropora* colonies observed by HMS *Challenger* survive. The mass mortality of the sea urchin *D. antillarum* in 1983–1984 precipitated their calamitous decline by spurring algal growth across the reefs. Around the same time, new destructive pathogens emerged. All across the Caribbean, staghorn and elkhorn corals began to display sickly bands of white at their tips, which spread rapidly along their hedgerow-style branches.

"White band disease" strips the coral skeleton of its protective tissue. Within weeks, the waterborne plague reduces the thinner branches of *A. cervicornis* to rubble. The sturdier *A. palmata* remains in place after death—a literal ghost of itself—until leveled by an inevitable storm. Its once dominant population devastated, *Acropora* was added to the official list of endangered marine species in 2006. The reefs that the branching corals leave behind are smaller, simpler, and support only a fraction of the fish varieties that dazzled the eye of Henry Moseley.

Warming seas are one culprit. Massive, colonizing coral species like *Acropora* play with fire—they thrive close to their natural thermal limit. The day *Challenger* arrived at the Virgin Islands in early spring, 1873—during *Acropora*'s prime—the temperature taken at the water's surface was 76°F (24.4°C). The equivalent March surface temperature today is 2°F warmer. During the unprecedented tropical heat waves of 2014–2015, when sea surface temperatures reached the mid-80s°F, coral bleaching killed one-third of *A. palmata* colonies in the Florida Keys. A more acidic ocean, too, is deadly to *Acropora* because inorganic carbon is less available for production of their vast interlocking skeletons, and increased carbon dioxide depresses the ability of tiny coral gametes to fertilize, settle, and grow.

Ocean warming and coral disease are synergetic. Corals bleached by overheated water shed their mucus skin, rich in antibodies, making them more susceptible to infection. White band disease was first identified in the Virgin Islands in the late 1970s. Then, in 2005, the first Caribbean-wide bleaching event spurred a fresh outbreak of the white band and "white pox" diseases, killing 17 percent of the remaining elkhorn coral colonies in St. Thomas and surrounding islands in less than three months. Observed up close, mucus from a dying coral gives off a foul odor, clouding the crystal-clear water. One source of these *Acropora*-killing diseases is wastewater from the nearby shores crowded with tourist resorts. The human bacterium *Serratia marcescens* has been identified in dead elkhorn fragments, a rare example of reverse zoonosis—the direct infection of a marine invertebrate by humans. Vectors for transport of *S. marcescens* from tourists' bathrooms to the corals they pay a small fortune to admire include an herbivorous snail and a large, stinging worm familiar to the naturalists of

HMS *Challenger*. One of his assistants at St. Thomas brought this "fireworm" to Moseley, carrying it gingerly in front of him. A full foot in length, with menacing bristles along its sides, the ubiquitous Atlantic *Hermodice carunculata* is a voracious feeder on corals.

Moseley laid the worm specimen out on the table. He identified two sets of tiny eyes—front and rear—and a small slit mouth on its underside for feeding on coral branches. The body divided up into segments, dozens of them, like strands of a ship's rope. Under the microscope, the toxic beard-like bristles that gave the fireworm its reputation assumed a "beautiful asbestos-like whiteness." Weeks later, to Moseley's surprise, they scooped an even longer fireworm from the surface of the sea near Bermuda, and then two more on their return to Cape Verde. Evolution had definitely armed the *H. carunculata* for success.[1]

The trip due north from St. Thomas 900 miles to Bermuda brought them to the high limit of the tropics and to the last of the great west Atlantic reef systems that *Challenger* would encounter. Corals, like most marine creatures, are acutely temperature sensitive. At Bermuda, waters were too cool for the reef-building *Acropora* that dominated the shores of the Virgin Islands. The branching fire corals—*Millepora*—took their place and caught Moseley's attention as an inviting object of study.

Only three months into their voyage, his senior colleague Wyville Thomson had written several papers for publication in the new London journal *Nature*, fashionable for its dedication to Darwin and the cutting-edge field of evolution. Meanwhile, a junior scientist on staff, John Murray, had already made several stupendous breakthroughs in bottom sediment research during their recent Atlantic crossing. The unappetizing ooze Murray collected from the seabed had yielded up fundamental

1 McIntosh (1887) 12:25.

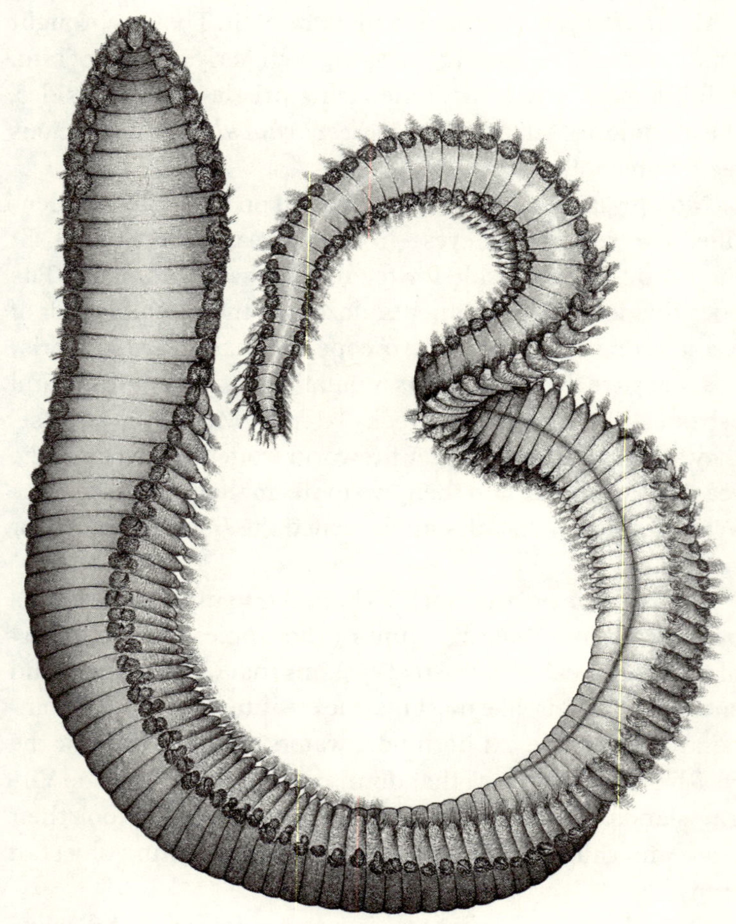

FIGURE 3.2 Bearded fireworm (*Hermodice carunculata*). *Scientific Results: Zoology*, vol. 12, plate 5 (1885).

secrets of water chemistry and the marine food web. Moseley had observed even young Willemoes-Suhm late at night writing up his notes to send to his famous mentor in Leipzig, Karl von Siebold, for publication in the most widely read natural science journal in Europe. To keep up with his ambitious colleagues, Moseley knew he needed an eye-catching research topic. What better than the *Millepora*, with their enigmatic tabular structure and affinity with well-known fossil forms? The fire coral might yet prove to be an ancient organism linked to the evolutionary past.

When he peered through his water glass in the clear lagoon off Bermuda, Moseley counted fewer coral species compared to the warmer Virgin Islands. The yellow "brain coral" dotted the sandy seafloor lit by sunshine, while others, such as the plate-like *Agaricia fragilis*, seemed to seek the protection of dim recesses and caverns in the reef. Altogether sparse.

The *Millepora*, by contrast, colonized every surface with their encrusting film, from glass bottles discarded in the harbor to fans of dead *Gorgonia*. The bright yellow branches of the fire coral were easy to locate but treacherous to handle, and no less difficult to study under the microscope. The soft, tiny polyps studding the corals' hard surface died instantly on contact with the air. And as for those specimens successfully taken on board and bottled, their polyps shrank from sight at the slightest movement of the ship.

Under Moseley's microscope on HMS *Challenger*, the *Millepora* revealed both their intricate beauty—and vulnerability. With assistance from Murray, Moseley placed lobes of fire coral in alcohol to encourage separation of hard and soft parts. Then, he carefully pried the living surface of the coral from its rump body, a green gelatinous mass he left in a jar to the side. With magnification, he saw that the surface layer of the fire coral consisted of "an irregular network of tortuous canals" full of round cells. He called these "nettle cells," tiny capsules filled with projectile poison threads designed to deter predators—and overcurious naturalists.

These cells, in turn, formed a protective zone around each individual polyp, which were themselves covered in what appeared to be little tentacles. The polyp was endowed with a mouth that in the living coral had a "conspicuous glistening white appearance," while its stomach contained sand grains

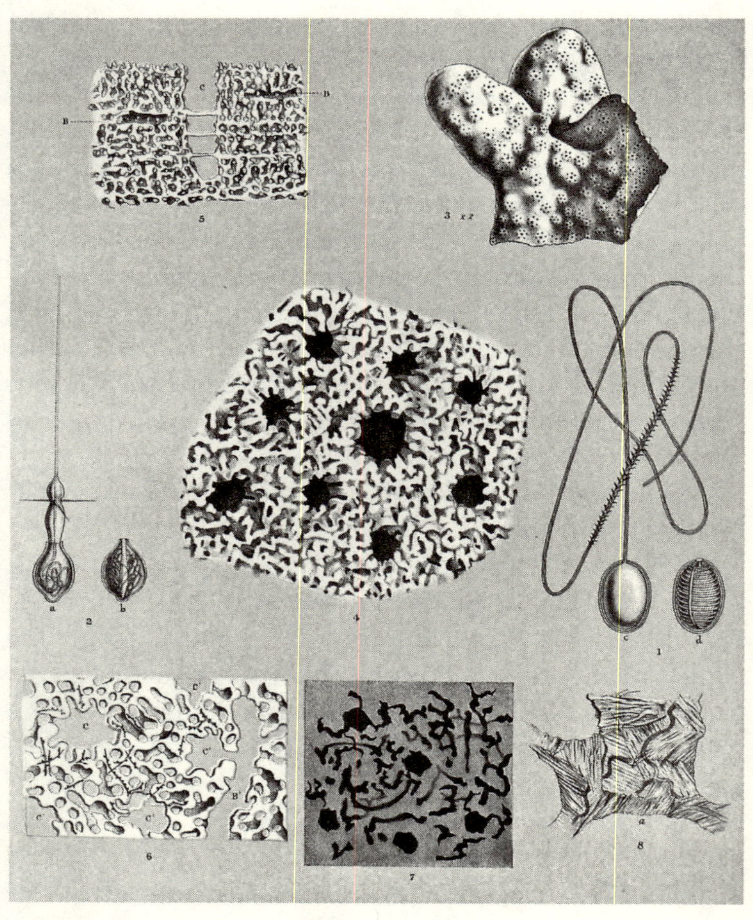

FIGURE 3.3 The interior structure of the fire coral, as first sketched by Henry Moseley. *Scientific Results: Zoology*, vol. 2, plate 13 (1881)

and sundry radiolaria. Most remarkable of all to Moseley, however, was evidence of vegetable parasites on *Millepora* similar to those he had seen on deep-sea corals. The fire corals of Bermuda—like those he would later collect in the Philippines and Tahiti—showed clear signs of infestation. Foreign organisms had bored through its soft surface tissue and into the calcareous heart of the coral itself.[2]

Henry Moseley was arguably the first scientist to identify potential weaknesses in Caribbean coral design. As warm tropical waters extend northward in the twenty-first century, *Millepora*'s vulnerable surface tissues are subject to unprecedented new stresses—and predators. During the summer heat waves of 1987, when deadly coral bleaching afflicted reefs throughout the Caribbean, the northerly situated Bermuda islands, bathed by cooler currents and winds, escaped unscathed. But the very next summer, the worldwide epidemic of coral bleaching reached Bermuda at last, laying waste to whole colonies of the fire coral *M. alcicornis*. When HMS *Challenger* dropped anchor at Bermuda in late April 1873, the surface water temperature was a cool, spring-like 67°F (19.5°C). In the disastrous summer of 1988, those temperatures breached 86°F (30°C) for the first time on record. White patches were first observed on outcrops of fire corals in early August. By summer's end, almost a third of Bermuda's reefs had turned completely white.

Equipped only with a low-power microscope, Moseley identified parasites on the Caribbean corals but not the vital codependent relationship between corals and the single-celled algae they host called zooxanthellae. These algae permeate the coral, forming microniches. In return, they provide energy to their coral host through photosynthesis and the colorful pigments that are its tropical trademark. In warming waters, however, the coral's respiration rate intensifies and with it its

2 Moseley (1876a) 113; Moseley (1881) 21.

demand for energy. If the algae cannot keep pace, the coral expels them, turning white in a persuasive pantomime of grief. During the worldwide extreme heat event of 2019, the *Millepora* reefs further south off the Brazilian coast began exhibiting telltale white patches as early as March. Not a single fire coral escaped bleaching, and only 10 percent have survived.

Several pathogens have been identified as agents of coral bleaching. One carrier of the coral-killing bacterium *Vibrio shiloi* is none other than the large, bearded fireworm collected by Henry Moseley—*Hermodice carunculata*. The fireworm wraps itself around an exposed coral branch, ingesting edible tissues from its tips. Historically, the impact of hungry fireworms has been balanced by the rapid growth of reef-building corals and the fact that numbers of reef fish include fireworms in their diet. But with corals weakened by higher sea temperatures and its natural predators depleted by overfishing, the fireworm now stands unopposed as it devours and infects vulnerable corals worldwide. In the process, it is unwittingly destroying the very reefs it depends on, as do millions of people in adjacent coastal communities. In the Anthropocene ocean, long-term, beneficial relationships turn toxic, and all creatures suffer, not least the small-bodied and juvenile fish who rely on the venomous fire coral for a safe place to hide or sleep and for abundant, ambient food. As color fades from our coral reefs, the fish they support, in their technicolor array, fade away with them.

That said, the Caribbean reefs as Moseley and Murray found them in 1873 were hardly pristine. Before Columbus, the tropical island coasts had been home to a menagerie of large sea creatures—including sea turtles, manatees, and an indigenous monk seal. By the time *Challenger* arrived at the Virgin Islands, the monk seal was extinct, and the turtle and manatee populations had been reduced to a small fraction of their original plenty—victims of the industrial-scale fisheries introduced by European colonization. For the reefs of the

Caribbean, the consequences of the disappearance of its large vertebrate predators are difficult to reconstruct. Suffice to say, the Victorian naturalists were operating within an already much-altered reef ecosystem. The Caribbean Sea had entered the Anthropocene centuries before.

Though bathed in tropical sunshine day after day, *Challenger* had made a gloomy passage from St. Thomas to Bermuda. At the outset, they lost one of their ship's boys to an unlucky accident with the dredge. To lift the company's spirits, the captain ordered a sail tethered over the side so the men, most unable to swim, could splash about and perhaps scrub themselves. Two boats, meanwhile, kept an eye out for cruising sharks.

They were sailing—slowly, with little wind—across the deepest waters they had yet sounded. To their amazement, they raised a small, bright red crab from three miles down where Forbes's azoic theory assured them no living creature could exist. On the evening of April 3, they caught their first glimpse of Bermuda's lighthouse, flashing red in the gloom. As if on cue, the weather turned cold and squally. For months, they had slept with their cabin windows open and no blankets. For dinner at the club in Bermuda, the naturalists traded their light jackets for tailcoats and vests.

The island chain was, at first sight, a disappointment: somber, low-lying, featureless. But once within the reef, a true coral paradise came into focus—a long, white beach fringed by curling surf and water of intensest blue that changed to green in the play of sun and shade. Tiny atolls dotted the harbor, with little sailboats darting to and fro between them crowded with fishermen.

When they rowed toward shore across the glassy surface of the lagoon, the reef loomed almost close enough to touch. A glorious palette of nature greeted them: vivid purples, greens, and reds of the branching *Millepora*, bulbous brain corals, and feathery *Gorgonia* crowded with sponges, seaweeds, and seagrasses of all kinds, in between strips of white coral sand. Fish swarmed all about, dominated by the large, reddish groupers. They spotted bright red starfish and an angel fish cruising sideways, flashing his coat of gold and blue.

The shore was a kind of bleached requiem for the reef itself, consisting of tiny grains of dead corals with compacted fragments of fish remains and mollusk shells by the billions. John Murray walked along the Bermuda beach barefoot in the manner of the bluejackets whose lifestyle he envied. He saw how the dry sand above low tide had over time been blown inland by the wind to form dunes fifty feet high. Over incalculable ages (to him), such dunes had formed the rocky foundations of Bermuda itself. These now supported fragrant, English-style gardens and cedar forests of "monotonous loveliness" but, where exposed, the mottled limestone, porous with tiny caves, still recalled the ocean floor from which it came long ago.[3]

The philosophers of HMS *Challenger* were all in conversation with the preeminent naturalist of the age, Charles Darwin. Some quietly believed themselves in competition with his radical ideas. Murray, for instance, found himself unpersuaded by Darwin's fashionable theory of coral atolls. He climbed the Bermuda dunes on the south shore to where the ever-encroaching sand bordered the island's red soil. Then he took a boat out to an atoll, no more than twenty feet wide, which he found dominated by calcareous tube worms topped with orange tentacles. Then he went back to the club to play billiards, drink beer, and think it over.

From observations aboard the *Beagle*, Darwin had drawn an elegant picture of atoll formation: subsiding volcanic islands left fringing reefs behind them like lost crowns of a vanished kingdom. Murray, by contrast, was convinced the reefs were built up from below atop the peaks of submarine mountains such as *Challenger* had regularly sounded in their

3 Campbell (1877) 18.

Atlantic crossing. The signature ring-shape of coral reefs, enclosing still-water lagoons, was due to the constant erosion of the reef by the action of the tide, which favored depletion of the middle.

Darwin's atoll theory has rarely been questioned since its first publication in 1842, which is odd considering subsidence is too slow a geological process to play a meaningful role in the formation of island reefs. Neither Murray nor Darwin held the key to how coral atolls are actually made—via rapid fluctuations of sea level driven by glacial cycles.[4]

Regularly over the last 2.5 million years, long-term ice ages have exposed flat banks of land and sand along tropical coasts. When flooding waters return during interglacial periods such as our own, fringing corals grow upward at the margins more rapidly than from the sunken center. Reefs, in short, are always in a race against sea level—a race they are fated to lose in the Anthropocene. The meter or more of global sea level rise predicted for 2100 will swallow up whole portions of Bermuda, beginning with her beaches, and will drown her reefs.

The death of the ship's boy had come when *Challenger*, replenished with coal, stores, and water after her Caribbean sojourn, paused to dredge the pure white ocean bottom north of St. Thomas. The hands, accustomed to the routine from tedious weeks on the Atlantic crossing, secured the lead line to the deck via the main yard. The "donkey" engine commenced its familiar roar.

But a loose hook on the *Challenger* deck had escaped attention. At a critical moment, the ship began to drift. The ropes groaned. Just as the dredge appeared in a cascade of foam, the iron block securing the lead line to the deck broke free, flying upward with the velocity of a cannonball. William Stokes, a ship's boy, happened to be passing the very spot. The airborne

4 See Droxler and Jorry (2021).

iron drove him against the side, crushing his leg. His left hand was "smashed to atoms" and blood flowed from a deep gash near his temple. They rushed him below, head lolling. But the ship's doctor, Alexander Crosbie, could do nothing for the unconscious boy. Sometime after midnight, his breaths became short and rapid, then ceased.[5]

Next morning, they found themselves floating atop a yawning abyss. After tacking oceanward through the night, *Challenger* had furled her sails before dawn. Captain Nares gave the order, and the steam engine came to noisy life. The sounding rod was lowered, and unused coils of hemp rope snaked across the deck before the rope touched bottom, nearly 4,000 fathoms beneath *Challenger*'s bobbing hull. They repeated the entire laborious procedure just to make sure. No temperature at that mind-boggling depth was recorded: both thermometers were ground to smithereens. Nor was anyone surprised when later the dredge collected nothing but cold red clay. After all, what living creature could survive under the weight of four and half miles of ocean?

In the failing light of evening, a solemn bell brought the entire ship's company on deck: scientists, officers, and men. The body of teenage William Stokes lay wrapped in a hammock, weighed down with cannon shot. Captain Nares read the familiar service. The words "We commit his body to the deep" rang with new meaning after the day's sounding, the deepest in the known world.

It would have been ghoulish to peer over the side to watch poor Stokes plunge into the Atlantic abyss. But the men tracked his watery descent in their minds. The next day, a delegation of bluejackets appeared at the door of the naturalists' mess with pressing questions: Would William Stokes sink directly to the bottom? If he did, what would his condition be after a four-mile downwards journey under pressure of water

5 Anon. *Journal* [March 25, 1873].

that could crush thick glass? Or would he not sink so far? Would William instead "find his level" where the density of the water matched his body, and float there suspended until the day the sea gave up its dead?

No one mentioned the boy's more likely fate in shark-infested waters.

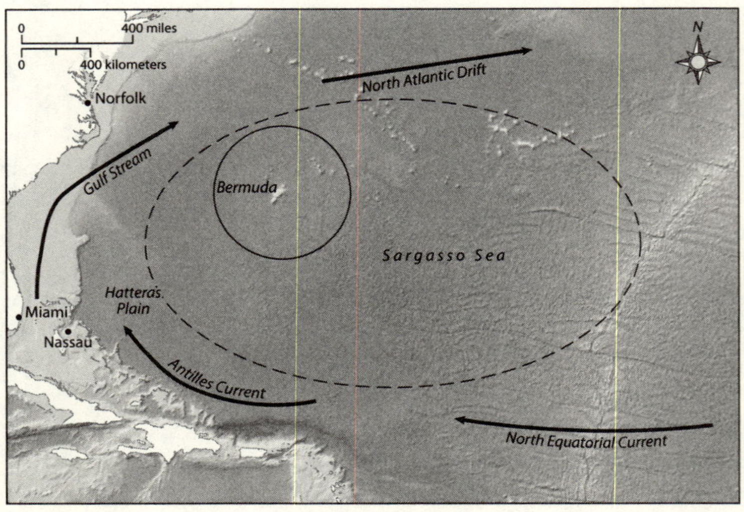

FIGURE 4.1 (Top) Sargasso weed (*Sargassum natans*). William Harvey, *Phycologia Britannica, or, A History of British Sea-Weeds*, vol. 1, plate 109 (1846). (Bottom) The Sargasso Sea, a creation of encircling currents and host to a unique marine ecosystem. Adapted from Struan R. Smith and Tammy Warren, "Bermuda and the Sargasso Sea," in *World Seas: An Environmental Evaluation*, 2nd ed., vol. 1, *Europe, the Americas and West Africa*, ed. Charles Sheppard (Elsevier, 2019).

CHAPTER 4

Wide Sargasso Seaweed

West Atlantic Ocean
20°35′ N, 40°70′ W

John Murray had joined the *Challenger* expedition at the last minute when Wyville Thomson's chosen assistant withdrew. This happenstance changed Murray's life and the history of marine science. At that time he was thirty years old, without publications, reputation, or a degree. But he was known in Edinburgh, the hub of Victorian marine speculation. Fellow Scotsman Lord Kelvin—a dominant figure at the Royal Society—once heard the young man discuss phosphorescence intelligently. That was enough.

The good fortune worked both ways. Not only did Murray prove himself adept at all scientific duties aboard *Challenger*, but when Thomson died of nervous exhaustion in the middle of publishing the expedition's results—a Herculean task that took two decades—it was Murray who brought the project to completion, authoring the narrative and parts of five of the fifty volumes himself, and supplying gap funds as government commitments faltered.

Murray was broad-shouldered and energetic, with a gruff, no-nonsense style. He had a disdain for formality of all kinds, which extended to his education. In his years at the University of Edinburgh he never once showed up to an exam. He enjoyed roughing it and had travelled to the Arctic on board a whaler. His scientific colleagues aboard *Challenger* were, without exception, "well born." It was left to Murray to venture below deck and fraternize with the ordinary seamen. He learned their songs and jokes and sympathized with their perennial complaints because he had suffered them himself, if only as a lark. The sailor's lot, he knew, was not a happy one.

Back in England, popular opinion since the days of Lord Nelson had held up the seamen of the Royal Navy as avatars of national virtue: honest, brave, and cheerful in adversity. Sea air and sun, too, in constant supply, ensured the British sailor lived healthy and happy. But Murray knew better. HMS *Challenger* could politely be called snug. In fact, it measured a mere 200 feet long and 40 feet at its broadest point amidships. Squeezed together for long hours in fetid darkness below deck and sustained only by salted meat, stale bread, and alcohol, the men suffered chronic maladies of all kinds.

Their mental health was no better. The typical foremast hand aboard HMS *Challenger* was situationally vulnerable to depression from his "total estrangement from home ties," a monotonous routine, and the constant, petty harassment of his officers—all for a career without realistic possibility of advancement. Stiff punishment for shirkers ensured dogged compliance from the bulk of the men. But without the presence of armed marines on deck—*Challenger*'s onboard police force—it was doubtful the officers could maintain control for five minutes. These officers, too, mostly rued their fate. Their families had nominated the Royal Navy in loco parentis from a young age, disposing of their surplus sons cheaply in the time-honored fashion.[1]

Steam power—an epoch-making technological advance—had arguably made seagoing labor conditions worse. *Challenger* herself represented an awkward compromise between sail and steam. She possessed the standard three masts, but with a bulky funnel protruding behind the foremast, spewing smoke. Thousands of tons of coal powered the ship during her three-and-a-half-year voyage around the world. Reams

1 Swire (1938) 48.

of black gold, plundered daily by Welsh miners in appalling conditions, were delivered to all corners of the Empire, piled high in dockside warehouses from Gibraltar to Hong Kong. The filthy, backbreaking business of loading sacks of coal from the dock into *Challenger*'s hold was mostly carried out by her nameless, underpaid sailors, not Murray and his fellow naturalists in their straw hats and white trousers, though they fully recognized its necessity.

Without an engine, the ship could not navigate to a precise location for sounding or steady herself for hours on end to operate the sounding lines and dredge. The gift of steam power allowed *Challenger* to assert her independence from contrary winds and currents, to chart a scientific course, and to stay on schedule. That said, a hold full of coal was good only for a few weeks steaming and must be husbanded carefully. Most of the time, *Challenger* operated as an old-fashioned sailing ship subject to the same freaks of weather that mariners had faced on open seas since the days of the Phoenicians.

The first days out of Bermuda were plain sailing; the fresh coal sacks lay untouched in the stifling hold. But then, as Captain Nares set their course north in the direction of New York, the wind shifted until it was directly in their faces and stiffened to a gale. Flocks of stormy petrels, true to their name, flapped among the rigging. Answering the captain's order to furl sails while the ship pitched and heaved, a foretopman named Whitfield fell from thirty feet aloft onto the hard deck. The next day, a monster swell struck the rudder, sending the ship's wheel spinning. The two experienced hands at the wheel knew well enough to let go and save themselves, but two boys under their instruction did not. They clung to it and were thrashed senseless on the deck. The same freak wave flooded the captain's cabin and floated his government-issued furniture out the door. Meanwhile, in the darkness of the engine room below, the shirtless stokers shovelled coal into boilers until the room temperature touched 130°F—all so *Challenger* could continue to press her course into the wind's teeth in defiance of the towering waves.

On the first of May they reentered the Gulf Stream—a bright blue river of warm water sixty miles wide and two thousand feet deep named for its source in the Gulf of Mexico and coursing headlong in the approximate direction of Great Britain. To Wyville Thomson and his fellow philosophers, the Gulf Stream was "the most glorious natural phenomenon on the face of the earth." The sounding line could not take the strain of the powerful current, broke, and was lost.[2]

Gulf Stream waters were 10°F warmer than the air above. A thick fog settled over *Challenger*, while below deck turned intolerably humid. Putting a hand anywhere on the ship left one's palm slick with damp; begin a letter home and the deafening rattle of the engine made ordering one's thoughts impossible. The misery of *Challenger*'s crew signified little to their captain, however, who concerned himself only with the risk of mass desertions at the next port. Eighty miles from Sandy Hook—gateway to New York—he dreaded the siren call of America to his men. So he altered course for dour Nova Scotia.

A day spent fishing for still-plentiful cod on the LaHave bank buoyed the company's spirits temporarily. But shore leave in Halifax gave full vent to the men's suppressed feral energies. Two were fined with drunkenness and assaulting police; more returned on board with black eyes and broken noses. Six of the ship's company never reported for duty at all, joining an ever-lengthening list of deserters from the greatest seagoing research expedition in history.[3]

The mood in Halifax during *Challenger*'s stay was solemn. A month before their arrival, the giant passenger steamer *Atlantic*

2 Thomson (1873a) 382.

3 The Atlantic cod fishery—engine of colonialism—finally collapsed in the 1990s after five centuries of exploitation. Cod stocks, which show no signs of recovery, are currently estimated at less than 1 percent of their precolonial volume.

had foundered in a storm off the coast. More than five hundred hopeful emigrants to North America drowned. Each morning, bodies that had washed up overnight were strewn like flotsam on the beaches. Posters on the walls offered anguished descriptions of the dead and a reward for their remains. The aurora borealis, which one night lit the entire northern sky in flame, seemed to cast a funereal glow across the scene.

Otherwise, the daily business of a Victorian Atlantic port town continued its relentless rhythm. The reeking fish market overflowed with cod, salmon, and halibut at cheap prices. A snapping lobster could be had for a penny. Stacks of fir and pine trees, hewed from the green forests behind Halifax, lined the dock alongside small mountains of coal. A hodgepodge flotilla of ships from the North Atlantic trade ranged along the wharf to gorge on it all. With *Challenger* depleted by her voyage in heavy weather from Bermuda, Captain Nares ordered up two hundred more tons of coal. Planning ahead, he also stowed enough tinned meat and biscuit to carry the ship across the Atlantic and back and from there into the icy wastes of the Southern Ocean.

"Dredge, sound, trawl; sound, trawl, dredge; and so on all the way," wrote one fed-up diarist of *Challenger*'s lower decks. Buried among the stores in the hold were several crates of sherry acquired in Madeira. These were designated "extra surveying stores" in the ship's manifest, a luxury rare in the Royal Navy. The sherry was compensation for the forty men required to operate the winches for sounding and dredging—ten hours or more of hard labor.

But the Navy's generosity only extended so far. When the equipment malfunctioned or the dredge came up empty, Captain Nares cancelled the sherry, as happened when the line gave way on the return journey to Bermuda—victim of the relentless coastal rip. The Queen's birthday offered a timely boost to morale. The entire ship's company was awarded a one-third pint from extra stores to mark the occasion. *Challenger*'s motley brass band played while, in a rare, democratic congregation, the men, officers, and philosophers smoked together on deck and toasted Her Majesty.[4]

When the temperature of the water climbed 20°F in mere hours, they had resumed the Gulf Stream. The cold, green waters of the Labrador Current gave way to a vivid indigo. Humidity enveloped the ship, heavy rain and squalls tossed them about, and the constant, grinding roar of the engine's propeller made conditions below deck "almost unbearable." They dropped buoys to measure the power of the Stream, then boats to sample it at different depths. The data showed that their single, rickety engine stood opposed to a hemispheric current delivering trillions of tons of warm water a day to the northeast Atlantic at a speed of four knots.

One especially tedious day, working into both the Gulf Stream and a howling gale, the ship made only forty miles under full steam. The men crowded the sick bay with fevers and rheumatic complaints. Three-quarters of the coal taken on board in Halifax was already gone. Nearing Bermuda, the captain gave the order to make sail, and authorized a welcome return to light duck trousers and shirts. Then the wind died altogether, and the ship's company broiled once more while the sails flapped limply like dying fish against the mast. A golden seaweed stretching for miles had surrounded the ship, clogging the rudder. After weeks battling the titanic Gulf Stream, they were now languishing in an "immense whirlpool" clothed in vegetation—the notorious Sargasso Sea.[5]

Deep in the bowels of HMS *Challenger*, her four boilers, charged with coal, condensed steam to water, releasing sufficient energy to drive the massive twin-blade propellor at her stern. In its essential mechanics then, the ship's motive force

4 Swire (1938) 44.
5 Matkin (1992) 73; Wild (1878) 73.

mirrored, in miniature, the planetary operations of the sea and sky whose undulating border she passed between as if pulled by an invisible thread.

The heat engine of the Atlantic Ocean is most intense at the Equator where volumes of evaporated ocean water rise up, driving the air at the top of the atmosphere poleward. These high-altitude airstreams, like the ocean currents beneath them, cool as they travel and sink in the approximate latitude of Bermuda (in the Northern Hemisphere). Even then their journey is not complete. The same equatorial low pressure that first drew the air column upward now sucks it back southward across the surface of the sea. These winds would blow due south were it not for the rotation of the Earth, which deflects them to the right, creating the trade winds that had ferried ambitious Europeans to the Americas for nearly four centuries before Captain Nares and his *Challenger* crew.

The airstream that filled *Challenger*'s sails in her Atlantic transit, when speeding across the ocean's surface, creates a frictional drag on the water, driving the currents that directed her hull. Columbus was the first to notice that his Caribbean-bound ship drifted westward in a calm—beneficiary of the North Equatorial Current. When that current confronts the great land mass of the Americas, it banks northward and joins the Gulf Stream, all part of a single oceanic system—a hemispheric gyre of mighty currents circulating clockwise along the coasts of Europe, Africa, and the Americas, with the Equator as its landless southern limit. The great North Atlantic Ocean gyre is not a simple, single circulation of water but contains complex branches and subbranches, one of which, between Bermuda and the Azores in the mid-Atlantic, HMS *Challenger* tracked eastward during the spring of 1873.

The accidents of Atlantic circulation at this latitude have, since the most recent breakup of the continents millions of years ago, created the world's only sea-inside-a-sea—a mid-ocean gyre that Columbus's men called Sargasso after the flowering algae carpeting its surface. The crew of the *Santa Maria* associated the presence of seaweed with shallow waters and were terrified of running aground. A full thousand miles across, the Sargasso Sea sits in the latitudes between the Equator and the trade winds known proverbially as the doldrums. After three days of pure, languid calm, Columbus confided to his diary a fear that these strange, windless waters would never release his ship.

It was the same inherited seafaring anxiety that had brought an embassy of *Challenger*'s men to the scientists' mess north of St. Thomas to inquire if young William Stokes's body, like the legion of men and ships lost before him in this sea of spells, would be trapped underwater forever, twisting in the watery weed. Jules Verne—whose *Twenty Thousand Leagues under the Sea* was published on the eve of *Challenger*'s departure—had Captain Nemo steer well clear of the "herbaceous mass" of the Sargasso Sea, though not before pausing to wonder at this "cold, quiet, immovable ocean . . . a perfect meadow . . . so thick and so compact that a vessel could hardly tear its way through it."

True to its enigmatic reputation, the Sargasso Sea shifts with the seasonal drift of the Atlantic currents that sequester it, so its borders can never be precisely defined. The *Sargassum* weed, too, can be elusive, regenerating itself in situ and capable of drifting near and far. *Sargassum* is the genus of many species. In circumstances unknown, *Sargassum natans* evolved 40 million years ago from the fixed, bottom-dwelling seaweeds ubiquitous along Earth's shores to a floating, open-ocean surface algae unique in the world. Over time, a diverse menagerie of creatures—barnacles, birds, crabs, mollusks, worms, insects, fish—was drawn to this floating prairie.

For weeks during their first westward transit, the *Challenger* philosophers had contemplated an empty or near-empty dredge drawn from clear blue waters barren of obvious life. Days passed likewise without sight of a single ocean wayfarer—no shark, or turtle, or dolphin, or whale. Then one rainy morning, 800 miles from any shore, a small garden of "gulf weed" floated by the ship. The philosophers crowded into the boats and scooped up everything within reach. Adding to the weird sensation of navigating across a vegetated plain, their soundings revealed that beneath the warm, weedy surface, the water column extended, clear and cold, for three full miles to the distant floor. That abyssal plain is now named for George Nares, *Challenger*'s captain, the first to sound its depths.

The Sargasso Sea was famous among Victorian mariners and naturalists alike. Henry Moseley had studied specimens in British collections, but these were dark and shrivelled and did not prepare him for the bright golden colors of the living weed. The *Sargassum* bobbed lightly on the water's surface, a single, feathery layer, with individual strands three or four feet long and loosely entangled.

At closer quarters, Moseley could see that the wind had divided the *Sargassum* into rows. The blue water in between struck a vivid contrast with the shining weed. Individual specimens consisted of a branching stem studded with short stalks and what looked like little berries—these were, on inspection, tiny bladders for flotation. A pretty white polyp covered the bladder like a net, heavy enough to break off. The water surrounding the weeds was full of these little white balls bobbing alone. Green, willow-like leaves—bursting with chlorophyll the same as their terrestrial kin—sprouted toward the end of the weed stem, ornamented with tiny flowering plants. The effect was very like a clambering vine on a garden bed except for its constant undulating motion and sparkling reflections from the water beneath.

For a cohort of Victorian scientists in the front line of public debates over evolution, these floating acres of gulf weed represented an ideal Darwinian laboratory. The creatures of the *Sargassum*, fully exposed to predators flying above them and swimming below and with only a scanty layer of weed for cover, had evolved to blend uncannily with their weedy habitat. The indigenous shrimps and crabs swarming over the *Sargassum* bed duplicated its specific shade of yellow in their shells, and were even blotted with irregular white spots to mimic the ubiquitous polyps and barnacles adhered at random to the weed's surface. The native Sargassum pipefish (*Syngnathus pelagicus*) had likewise developed skin tags indistinguishable in shape and color from a seaweed frond.

Moseley found worms and mollusks with a similar camouflage, indistinguishable from the weed itself until he got really close. One short-tailed crab—*Nautilograpsus minutus*—appeared to have the power to adapt its color markings to whatever specific bed of seaweed it inhabited, and was a tenacious resident. Sitting beside Moseley in the boat, Murray amused himself by plucking this crab from the weed and setting it on the open water. When the crab swam immediately back to its floating home, Murray tried again with the same result. The philosophers tired of the game before the crab did.

A little frogfish drew their particular attention, an animal they recognized from English beaches. The Antennariidae, a fish family renowned for its ugliness, possess a head and jaw disproportionate to the body, and leprous spots. This Sargasso variety, however, *Histrio histrio*, had adapted to its pelagic existence with a set of extra forelimbs—long, arm-like fins with which to grapple its host weed. Secreting a gelatinous string to bind strands of weed together, the frogfish made its nest out of *Sargassum* itself. Inside one such floating globular mass, they discovered a deposit of tiny frogfish eggs.

From the masthead south of Bermuda, in the heart of the Sargasso Sea, they could see the golden weed stretching in all directions amid blooms of blue-green algae—a banquet for marine life in a sterile ocean. The wind dropped for days until the mere appearance of "cat's paws"—a ripple of breeze on the motionless water—was enough to create excitement on deck. But as the *Challenger*'s progress slowed to nothing, their some-

FIGURE 4.2 Sargassum frogfish (*Histrio histrio*). Marcus Elieser Bloch, *Ichthyologie, ou Histoire Naturelle des Poissons*, vol. 1, plate 111 (1796).

time marine companions, not bound to wind or steam power, flocked to the scene. Sharks, turtles, and porpoises, with shoals of barracuda and swordfish, made a constant traffic among the *Sargassum* weed, giving it the appearance of a high-seas playground, or more accurately a crossroads, drawing travellers of all kind to feed, breed, and refuel on their way to distant coasts unknown.

In a normal year, a million metric tons of Sargasso weed drifts westward from the Caribbean Sea into the Gulf of Mexico, then northward along the eastern seaboard of the United States via the Gulf Stream. The majority of that massive tonnage is then captured by powerful eddies of water spinning from the mid-ocean gyre comprehending the Sargasso Sea—a closed, self-sufficient system. The whirlpool energy of the gyre is so powerful that the same water might circulate in the Sargasso Sea for fifty years before its slow release to the greater ocean. On a different timescale altogether, the indigenous Sargasso fauna likely evolved in its embrace millions of years ago—in a sea without borders and seemingly outside of time itself.

Few of the lifelong residents of the Sargasso Sea—mollusks, crustaceans, worms, hydroids—actually eat the weed. Rather, the *Sargassum* releases nutrients into the surrounding water, providing a first link in the food chain. Its grateful tenants return nutrients to it and migrate daily through the water column, thereby connecting their surface community with the denizens of the deep sea.

Others make the Sargasso Sea their nursery or temporary haven. Baby green and loggerhead turtles swim hundreds of miles from the beaches of their birth to the Sargasso Sea while, most remarkably of all, the European eel crosses an entire hemisphere to lay its eggs by the millions in the shelter of the *Sargassum* weed (both turtles and eels are now critically endangered).

Centuries of superstition and legend surround the Sargasso Sea. But even stranger things are now afoot. The *Sargassum* is expanding. The winds that bullied the *Challenger* on her journey along the American coast and east to the Azores are the same that perpetuate the Sargasso Sea and drive circulation of its signature weed. The sea sits at the heart of the wind rotation formed by the prevailing westerlies and northeast trade winds in the temperate west Atlantic, which themselves shift a few degrees north or south according to largely predictable cycles.

But in the winter of 2009–2010, this so-called North Atlantic Oscillation presented an "extreme wind anomaly." Unusually strong westerly winds pushed southward well beyond their habitual limits for several months in spring, driving large floating mats of Sargasso seaweed with them into tropical waters and from there to all corners of the Atlantic. *Sargassum* was reported for the first time off Gibraltar in 2010, while in 2011 great swaths of weed clogged the coastal fishing grounds of West Africa from Sierra Leone to Nigeria.

Meanwhile, back in the western Atlantic, an unprecedented "golden tide" of *Sargassum* found its way into the Caribbean

Sea and Gulf of Mexico, washing up in piles ten feet high along beaches filled with tourists. Over the past decade, *Sargassum* has not only extended in range but also grown exponentially in size. The sheer amount of floating weed quadrupled in 2015. By 2018, it had fully burst the bounds of its eponymous sea to form a "Great Atlantic *Sargassum* Belt" stretching from Africa to the Americas. The inflated *Sargassum* biomass now measures 20 million metric tons, twenty times its historical average.[6]

The emergence of the Great Atlantic *Sargassum* Belt has exploded the long-held belief in the Sargasso Sea as a closed system. It raises too the specter of epic oceanic change, a tipping point reached where the altered winds, currents, and toxic runoff of a superheated west Atlantic Ocean now operate under a new regime with different rules for flora, fauna, and humans alike.

Scientists are divided on the nutrient sources for the metastasizing *Sargassum* Belt. Because the pelagic weed is capable of vegetative growth by simple fragmentation, new ocean waters assisted by rising temperatures might be sufficient to sustain fresh, expansive colonies of *Sargassum*. Others point to increased deforestation, flooding, and fertilizer use in regions of the Congo River in Africa and of the Amazon and Orinoco deltas in South America. Artificially high levels of nitrogen issuing from the mouths of these great rivers could theoretically supply and sustain a new "tropical Sargasso Sea."

Certainly, bigger is by no means better in the case of the Sargasso Sea, at least for human coastal communities. In the 2018 *Sargassum* beaching event in Mexico, stranded rotting weed fed algal blooms hundreds of meters offshore. Coastal waters pumped with ammonium and hydrogen sulfide sparked mass mortality in local fish and crustacean populations, killed off shoreline corals, and suffocated native seagrasses. Once

6 Johns et al. (2020) 17.

clear blue waters are now perennially turgid with no near-term prospect of recovery. The overpowering smell of rotten egg gas threatens to empty beaches of sunbathers and has pushed the regional tourism industry to the brink. More serious still is the very real prospect of arsenic contamination of Caribbean and Gulf beaches from rotting seaweed—a veritable "silent spring" for shoreline creatures throughout the tropics.

HMS *Challenger*—oblivious to this weedy future—enjoyed a rare spell of bright weather in the Sargasso Sea en route to the Azores. Between the vast fields of golden *Sargassum*, the open ocean resumed its clear blue perfection. With little of interest in the trawl, the philosophers contemplated the shifting colors on the water's surface as it passed from sun to shadow, flecked with white crests in the light breeze. Except for the slight heel to port, they felt no sense of motion on deck though *Challenger* was cruising at a clean ten knots. On such days, strung together, it seemed they might sail forever across this flawless element, the bright line of the horizon dipping perpetually five miles beyond the bow.

Then the wind dropped and they fell again among the *Sargassum* weed, teeming with life. Out in the boats, they scooped up rare white starfish, tiny crabs never seen before, and a host of tiny insects on parade across the surface. They rescued a sinking turtle with fins so encrusted with barnacles it could no longer swim. Willemoes-Suhm brought up a new, crimson-colored shrimp that suggested a link between two established species. He hurried off to dissect and describe the find for his debut publication in England.

Then, on their very last day in those ever-surprising waters, they fished an enormous translucent *Pyrosoma*, five feet long. No one had ever seen a tunicate so large. The entire surface of the translucent, tube-like creature was covered in pink spots; when they laid it in the bath tub, it lit up like a lantern. Kneeling alongside, the philosophers inscribed their names with their fingers on the creature's body, each letter perfectly legible in the falling darkness. Hopes were high for a famous trophy from the Sargasso Sea, but next morning the *Pyrosoma* had disintegrated in the bath into a thousand tiny, colorless lumps. The spectacular tunicate is in fact a zooid colony, whose members had disbanded under stress.

That day's sounding—June 25, 1873—showed that the surface layer of warm water that had bathed the *Challenger*'s hull since Bermuda was thinning out, soon to vanish entirely. Within hours the current turned abruptly southward, and all trace of floating golden seaweed disappeared. They were surrounded once more by the familiar Portuguese man o' war, bladders aloft. The surface water temperature dropped still further, and a cold wind from the north brought squalls and heavy rain. They had exited the Sargasso Sea.

In the months that followed, *Challenger* sailed east back across the Atlantic to Tenerife. Captain Nares then turned the ship south along the coast of Africa, and from there westward again and a second transatlantic haul, this time on a southerly route to Brazil. In the minds of the Challengers, they had left the Sargasso Sea well behind. But they had not reckoned on the phantom power of the sea-within-a-sea to remake itself. Sailing east to west across the Atlantic at a tropical latitude in long ago 1873, *Challenger* unwittingly traced the course of the "Great Atlantic *Sargassum* Belt" of our twenty-first-century Anthropocene ocean—a strange new chapter in the hoary legend of the Sargasso Sea, one undreamt of by Captain Nemo.

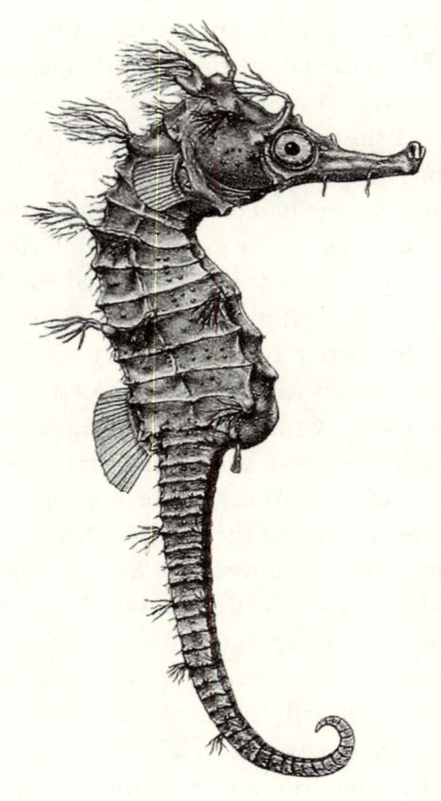

FIGURE 5.1 *Hippocampus erectus*. From *Scientific Results: Zoology*, vol. 1, plate 1 (1880). (Inset) HMS *Challenger* Legs 4 and 5. Halifax to Cape Verde to Bahia, Brazil to Cape Town. May–October 1873.

CHAPTER 5

Darkness Visible

False Bay, South Africa
34°11′ S, 18°26′ E

During her first full year at sea, HMS *Challenger* crisscrossed the Atlantic Ocean no fewer than four times. The ship did not adhere to tried-and-true routes favored by the hundreds of vessels ferrying goods and people between the great Atlantic ports, but crossed each time at an original latitude: first, from the Canary Islands to St. Thomas at roughly 20 degrees north; then, beginning in June, a more northerly return transit from Bermuda via the Azores to Madeira.

A summer outbreak of smallpox on Madeira ruined their chance to revisit the island. So they hugged the African coast for two thousand miles before tacking west near the Equator. This was *Challenger*'s shortest Atlantic commute, where the tectonic puzzle pieces of Africa and South America last joined 160 million years ago—when lush Brazilian forests were arid sands stretching west of modern Nigeria.

At the end of September, a case of yellow fever below deck drove the Challengers southeast from sweltering Bahia into the cold, deserted sea lanes of the South Atlantic. There they battled boredom and a steepling swell, rescued two marooned German sealers from Inaccessible Island—a tiny, volcanic pinprick of land in the ocean vastness—before coming to anchor at last along the sheltered tip of Africa, near Cape Town. There, on December 7, 1873, *Challenger* celebrated her first anniversary as the world's pioneer floating marine academy.

Freed from the monotonous routine of their latest Atlantic crossing, the officers and men crowded the boats for leave onshore. Some nurtured plans to desert the *Challenger* and make their fortunes at the new diamond mines inland. The naturalists, however, remained on board, where their most important work had only just begun. Hundreds of bottled specimens—from dozens of deep-sea stations across the Atlantic—crowded the tables of the workroom. Thomson and Moseley—with the assistance of ship's boy Frederick Pearcey and a manservant named William Pembre hired in Bermuda—replenished the specimens with fresh alcohol from giant casks stored overhead. They then passed each tubular jar carefully to Murray, who packed them into crates for their six-thousand-mile return journey to Britain—for future dissection and, ultimately, publication to the world.[1]

Beneath the ship's lilting hull, currents of warm and cold water—from the disputing Atlantic and Indian Oceans—alternated with the shifting wind. The summit above the bay offered fine northerly views of Cape Town and Table Mountain as well as the cape itself to the south, gateway to the forbidding polar ocean—*Challenger*'s destination after refitting. Beyond the horizon, ships rounding the cape for Asian ports and the Australian colonies made a steady, fast-moving traffic. Closer to shore, those heading west on the return journey to Europe stood within view of a telescope—and presented a pitiable contrast. They labored directly into "the great surface-current of the Southern Ocean," uninterrupted by land in

1 Pembre served as Wyville Thomson's personal servant until the stay at Hong Kong, where he died of undocumented causes. His appearance in a group photograph is his sole official recognition, an uncredited fate characteristic of many hundreds of workers, onboard and in ports en route, without whom the *Challenger* expedition could not have happened. Fifteen-year-old Frederick Pearcey, meanwhile, seized his opportunity as a workroom assistant onboard, and continued in that role in Edinburgh after *Challenger*'s journey ended, during the yearslong project of sorting and arranging specimens.

its course around the 60-degree parallel—the wildest waters on Earth.[2]

At the cape, *Challenger*'s senior scientists repressed their desire to explore and instead pulled extra shifts out of consideration for their young colleague Willemoes-Suhm, who ferried select crustaceans back to his cramped cabin to compose reports for the Royal Society. The German zoologist's urgent tasks included correcting errors that had marred his debut appearance in the journal *Nature* back in March. Adding to his anxieties, Director Thomson had prohibited him from publishing any of his *Challenger* research in German-language journals.

His compatriots recently rescued from Inaccessible Island—spared from ship duties—read Dickens by lantern light to improve their English. Willemoes-Suhm, meanwhile, sought relief from his work on the breeze-blown deck. He contemplated the sea ablaze with light, the work of ubiquitous dinoflagellates—identified by Murray as *Noctiluca* and *Pyrocystis*—and the tiny larvae of the bay's resident crustaceans, likewise gifted with mysterious, illuminative powers. Aristotle had been the first to describe the sea assuming its enigmatic nighttime glow, an object of fascination for all mariners. After twelve months cruising the Atlantic, bioluminescence appeared to Willemoes-Suhm and his fellow scientists as an almost universal trait of oceanic life.[3]

The man they called simply "Suhm" watched the fish swimming about the ship trail a shimmering light in their wake. When they banked or broke the surface, their stippled sides shone like silver. A question loomed over these brilliant motions: Why? In the past, naturalists contemplating a luminescent sea might have been satisfied with a tribute to the beauty of Creation. This

2 Wild (1877) 92.

3 An impression borne out by subsequent research: thousands of species of starfish, sea lillies, sea cucumbers, arrow worms, crustaceans, and deep-sea fish—and their larvae—deploy bioluminescence as a tool of survival.

was the default explanation for the colorful plumage of birds, for instance, and gaudy tropical fish. But Darwin's *Origin of Species*, published little more than a decade previously, had changed the terms of the debate entirely. Now when Suhm looked across the glitter-ball surface of False Bay, he speculated on how luminescence might advantage marine creatures in their struggle for survival. Was it to lure prey? Frighten an enemy? Or simply to light their way in a dark and murky ocean?

The silent fireworks on the water stretched from *Challenger*'s stern almost to shore. Within that penumbra of light, Suhm could clearly discern their compatriot vessel, HMS *Rattlesnake*, anchored off the port side. The warship, chockful of marines, had just returned from a bloody engagement in the Ashanti wars, fought intermittently during Victoria's reign to ensure unfettered British access to the African Gold Coast. Her commander, it was rumored, had taken a bullet to the chest and was confined to his cabin. But *Rattlesnake* would soon return. The Ashanti king, his palace reduced to ashes, capitulated within months.

Through November, Suhm labored on his research papers into the night with only a loquacious parrot, acquired in Brazil, for company. The May issue of *Nature* sat conspicuously on his desk. In it, Wyville Thomson had exercised his right as scientific director to publish details of Suhm's first significant discovery of the *Challenger* voyage two months earlier.

En route westward to the Virgin Islands, they had observed the first patches of gulf weed drifting along the hull. At night, the ship's wake lit up with animalcule flashes and the pulsing glow of jellyfish beneath the surface. Then one afternoon the dredge delivered, along with its usual deadweight sludge of red clay and calcareous foraminifera, a creature none of them had expected: an eyeless lobster, bright red. Suhm claimed the prize—an entirely original genus—and named it *Deidamia*, meaning "dauntless," his tribute to a blind crustacean struggling for existence on the cold ocean floor.

The philosophers had discussed the enigma of the blind lobster at their nightly seminar in *Challenger*'s library. The *Deidamia* possessed no eyes, no eye stalks, not even the remnant of an eye dwindled to uselessness in the permanent gloom of the deep sea. It was an adaptation, they guessed; but, if this were the case, why did other crustaceans from similar depths retain eyes of exaggerated size and sensitivity?

As they debated into the wee hours, the *Challenger*'s wake blazed with the combined luminosity of sea creatures of all sizes. The illumination reached even into the rigging, where the men working the sails appeared like ghosts eerily lit against the black sky. This nightly spectacle only deepened the mystery of Suhm's blind lobster. In a luminescent ocean where marine life generated its own light, wouldn't vision be an asset at all depths? How could an ocean floor predator, however dauntless, survive without it?

Then, in Halifax, a further complication. A museum curator from Buffalo, New York, had read the announcement in *Nature* and sent word to *Challenger* that Suhm had been scooped. The *Deidamia* name belonged already to a North American insect—the so-called sphinx moth, whose dark, mottled wings were cleverly designed to blend against the bark of boreal trees. Suhm and Thomson pondered this discouraging news and agreed he had no choice: a new name must be found and an awkward second announcement made.

Now on the far side of the Atlantic, Suhm contemplated his unlucky lobster once more, drained of color in a spirit jar. Two more species of the new crustacean genus had been brought up in the intervening months, tangled in the hemp tassels attached to the dredge: one in the West Indies, the other thousands of miles south along the coast of Inaccessible Island.

As if the enigma of blindness weren't enough, these specimens all bore an uncanny resemblance to a well-known extinct group of *seeing* lobster, the Eryontidae (Eryonidae). The fossil decapod *Eryon* possessed eye stalks but was otherwise a match with *Deidamia*. A tantalizing new theory surfaced in their discussions. Was this elegant lobster one of the longed-for "living fossils" *Challenger* had been charged to find? Did the incremental modification of its eyes over geological time prove the existence of natural selection? Suhm decided he would imply as much in his paper for the Royal Society but not state it outright. "It is very interesting," he ventured to suggest, "that cousins of the Jurassic Eryontidae are still living in the great depths."[4]

The young zoologist's faltering confidence had received a much needed boost on their arrival in Cape Town. The American curator who had disqualified use of *Deidamia* was now proposing the name *Willemoesia* for the new genus. Suhm was stunned at first by the news, then surprised by joy. To have his name attached to this celebrity lobster, a genuine living fossil, was like a brush with immortality.

Working late by candlelight in the calm waters of False Bay, Suhm's attention turned to another deep-red crustacean floating in its jar, which he had marked for description in his report to the Royal Society. It recalled a spell of blissful June weather west of the Azores when the island of Flores seemed to float at the horizon like a little cloud. Sounding in 1,000 fathoms, the dredge had delivered a magnificent, rare *Gnathophausia*, hitherto known only by a single specimen from the Indian Ocean. This novel shrimp, the size of his hand, sported extravagant red antennae and rows of feathery legs attached to its handsome crimson carapace. Like *Willemoesia*, the crustacean showed ocular oddities: in its case, twin dorsal "accessory eyes" additional to its main, stalk-mounted pair out front.[5]

4 Willemoes-Suhm (1876d) 577.

5 Willemoes-Suhm (1875) 35. When, a month later, the trawl delivered *Gnathophausia gigas* from the polar waters off the Kerguelen Islands, it was further evidence of the close relation between high-latitude fauna at both ends of the Earth. In the global distribution of species, the Challengers surmised, temperature transcended geography.

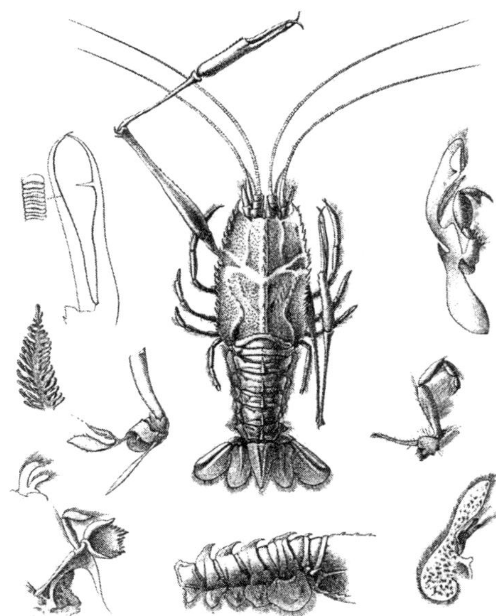

FIGURE 5.2 The blind lobster *Willemoesia leptodactyla*. From *Scientific Results: Zoology*, vol. 24, plate 18 (1888).

The shrimp *Gnathophausia* raised further questions about the nature of light and seeing in the ocean—and also fresh doubts. Were its supplementary cavities perhaps not eyes at all but organs of luminescence? Designed not for seeing but for illumination and for being seen? En route to Bahia, a new deep-sea fish presented the same tantalizing possibility: *Ipnops murrayi* possessed a long body, broad snout, and an "extraordinary apparatus on the upper side of the head." Moseley conjectured these must be eyes, albeit degenerated; but Murray thought it a "luminous organ" designed to emit light not receive it. Whatever purpose the cranial arrangement served, it demonstrated a remarkable adaptation to a lightless existence.[6]

Once Suhm had completed his descriptions, *Gnathophausia* followed *Willemoesia* and *I. murrayi* into the crate, bound

6 Günther (1881) 190. Bioluminescence, which the Challengers called "phosphoresence," has a host of applications in addition to camouflage, for both predators and prey. *Gnathophausia*, for example, shoots bright blue clouds from nozzles on either side of its head to distract predators.

for the desks of European authorities whose task it would be to reach a judgment on this and a host of other *Challenger* mysteries.

With its complement of fifty adolescents rated "boys," Captain Nares had been obliged to employ a schoolmaster aboard HMS *Challenger*. Mathematics formed the core of naval instruction, to fit young midshipmen for calculating the position of a moving ship out of view of land. But *Challenger* proved an unlucky ship for learning. Their first teacher had retired to his hammock in Bermuda one night never to rise again. The official postmortem settled on "apoplexy." Then in Cape Verde, where they were due to welcome his replacement, the new man had unaccountably disappeared on a morning walk so that *Challenger* was eventually forced to leave without him. The schoolmaster's body was found weeks later at the bottom of a cliff, the victim, it was supposed, of dehydration and an overambitious climb.

Hauling south out of Cape Verde in early August, *Challenger* embarked on her third consecutive crossing of the Atlantic. A burst of wet, squally weather came as a relief after the dust bowl of Cape Verde. Captain Nares skirted the African coast, driven by the Guinea current in quest of the southeast trade winds to propel them once again to the Americas. The Ashanti Kingdom, under siege by British troops, stood due east.

Since before their arrival at Cape Verde, *Challenger*'s proximity to the coast meant the blue ocean turned intermittently to green—a product of upwelling and the turbid, outflowing rivers beyond the eastern horizon. These plankton-rich waters, unfurling like tongues from the coastal delta into the sea, inspired the islands' name, as well as a nightly carnival of bioluminescence more spectacular than any they had yet seen.

First, the familiar sea-pickle *Pyrosoma* carpeted the sea for miles around with a blue-green glow. Then, at intervals, "momentary scintillations" of crustacean larvae lit up the surface with lilting mats of dinoflagellates sparkling in their millions. In his diary, Moseley called the effect "beautiful" and "weird." Looking down from the bridge, he watched the light burst out where the *Challenger*'s side cleaved the surface, enough to illuminate the sails flapping above his head. Edging nearer the coast, the nighttime swell took fire. Billions of Murray's *Pyrocystis*—single-celled but visible to the naked eye in a jar—gave the surface a pulsing, diamond effervescence. To the rear, *Challenger*'s wake quivered like a great candle flame, while ahead the ship's plunging bow created sheets of colored light that hung for a long moment suspended against the blackness, flashing light on the faces on deck "like so many spectres."[7]

The tow net was in constant operation along this remarkable coast, in addition to the trawl. Then from the surface one evening a fresh surprise: a perfectly transparent amphipod, its working organs plain to see within its clear, gelatinous body. From the ocean bottom, meanwhile, an equally see-through crustacean emerged, apparently new to science, which the professor immediately handed over to Suhm. They had seen this remarkable shrimp—which he dubbed *Thaumops pellucida*—once before, off Portugal. In the succeeding days along the southwest coast of Sierra Leone, they met with several more. At night, the surface tow net snared one, then during the day more glassy crustaceans emerged from abyssal depths via the trawl.

The circumstances of *T. pellucida* raised a new debate among the scientific staff. The enigma of translucency aside, was its capture at radically different depths evidence of a great daily migration? Did the inhabitants of the deep rise collectively to the surface at night to set the sea alight before retreating ahead of the sun's rise? And if so, why expend such enormous effort to make the journey?

7 Moseley (1892) 498; Murray (1895) 1:316–18; Anon., *Journal* [August 17, 1873].

Suhm's impatience for professional renown—and the lack of a full library onboard—now led him astray once more. At Bahia, during their Brazilian sojourn, he had published his discovery of *Thaumops pellucida* only to be obliged to retract his claim on arrival in Cape Town. A Frenchman named Guérin-Méneville had already described the species—he called it *Cystisoma neptunus*—from a partial specimen retrieved in the Indian Ocean. That his French rival's effort amounted to a paltry few lines and a bad drawing did not erase his priority. Stewing over the mistake in Cape Town, Suhm had an inspiration. Perhaps if he trumpeted the uniqueness of his shrimp-like amphipod in this new paper, he might yet upgrade the *Cystisoma* genus of Guérin-Méneville to a family, call it the Cystisomidae, and claim that for himself.

A sea lit like fire. Fish with lamps for eyes. Crustaceans the color of royal robes. Transparent shrimps. A deep-sea odyssey en masse to the surface and back—the largest animal migration on Earth—enacted daily. Tacking back and forth across the Atlantic, pausing every second day to sound and dredge, the *Challenger* scientists had enjoyed the luxury of a full year to ponder these wonders. Some enigmas they solved; others they left to future generations of oceangoing philosophers.

Wyville Thomson was first to suggest that animals in the deep sea sourced their own light. In wild, luminous waters off the African coast, it was impossible to escape the impression of an ocean world governed by a different sun—the collective energy of legions of its creatures. Recent research has established that the eyes of deep-sea fish are a hundred times more sensitive than the human eye. For the *Challenger* scientists, the laws of terrestrial light—and seeing—seemed hardly to apply to the lurid, starry medium of the sea. It was a reminder that the human experience of nature's wonders is not simply a product of external biological reality, but an artefact of seeing itself.[8]

8 Warrant and Locket (2004) 692. The German naturalist Robert von Lendenfeld, tasked by Thomson with speculating on the function of phosphorescence

Even at sunny noon in False Bay, Suhm's simple, camera-styled eyes were ill-equipped to see as ocean creatures saw, or to see them at all. When he peered over the side rail, the sea answered his gaze most often with no more than a shudder of its meaningless bulk, a glare of light, or a mere blue-toned reflection of the sky above. The entire scene was a smiling rebuff to human curiosity. To describe the brilliant exuberance of the deep sea from this distance, or from the contents of their meager nets, was like imagining the pyramids from a handful of desert sand.

One hundred fifty years after HMS *Challenger*, the science of bioluminescence is still in its infancy, though the sheer strangeness of the deep ocean has come into better focus. Unlike a terrestrial environment—with its forests, grasslands, and rocky hills—the open sea, miles deep, offers no refuge for its animal residents. In shallower waters near the coast, teeming coral gardens and kelp forests offer abundant cover. But beyond this haven, across the vast majority of our blue planet, there is literally nowhere to hide. For thousands of marine species—from microorganisms to full-sized fish—survival depends on alternative forms of self-protection, most notably camouflage.

Camouflage itself takes a variety of wonderful forms, including bioluminescence. Pelagic shrimp, for example, practice a form of self-concealment called counter illumination. Most deep-sea predators are endowed with upward-turning eyes to scout for prey feeding at the nutrient-rich surface. To elude them, light-emitting organs—named photophores—on the shrimp's ventral side replicate precisely the color and intensity of the ambient sunlight (or moonlight) from above, rendering it invisible to its enemies swimming below. Daylight

---

in the oceans, agreed with him that "the large eyes of many deep-sea fishes [collected by *Challenger*] show that there must be some light in the depths inhabited by them, the source of which can only be sought for in phosphorescent organs." Lendenfeld (1887) 324.

filtered through the absorbent ocean becomes bluer and dimmer with depth. Because this light, scattered through the water, is ever-changing, and the creatures themselves are in constant motion, luminescent camouflage is not set-and-forget. Photophores alter brightness and angle second-by-second to match the lit surrounds and thereby maintain the animal's precious cloak of invisibility.

As students of Charles Darwin and Alfred Russel Wallace, the *Challenger* philosophers understood the principle of camouflage in nature but had no notion of counter illumination to explain the nightly spectacle of their ship bathed in glowing light. They speculated in vain, too, on the perennial question of why crustaceans are red. It seemed to Moseley a minor tragedy of nature that the majestic lobster *Willemoesia* should waste its gorgeous scarlet shell lurking in the darkness, never to be seen. It could only be "for some purpose" unknown—or a case of a persistent trait rendered useless over time.[9]

The redness of deep-sea crustaceans has since been connected to bioluminescence, also. Beyond several hundred meters, no downward-dwelling sunlight penetrates the ocean. Only luminous flashes punctuate the darkness, like searchlights scoping for predators or prey. Enter the red lobster. The bioluminescent spectrum being blue-green, redness is invisible beneath these lights, allowing the crustacean giant to hide in plain sight. This so-called cryptic coloration represents one of multiple evolutionary responses to the necessity for camouflage in the populated mid-waters of the open ocean.

In other instances, the work of camouflage was more obvious. During their meanderings in the gin-clear Sargasso Sea, Moseley had observed how residents of the floating weed—tiny shrimps and fish—had adopted the same golden color as their host, even mimicking the white lacework of *Membranipora*

9 Moseley (1877) 22.

encrusted upon it. And by the end of their Atlantic odyssey of 1873, evidence of yet another evolutionary strategy for self-concealment had become overwhelming. The ubiquitous glass squid (*Cranchia*) and Suhm's remarkable see-through amphipod *Cystisoma* were not freaks of nature but only extreme examples of transparency adopted for camouflage.

Unlike luminescence or cryptic coloration, transparency, to be successful, requires the commitment of the entire animal. In the little boat, scouring the surface with his net, Moseley captured weird, glassy creatures by accident that had been invisible to him inches from his face. Even when he held Suhm's little shrimp up to the light in a glass globe, he could barely make it out. The *Cystisoma*, it turns out, possesses antireflective devices on its body and legs (similar to butterflies) and a transparent retina stretched across its entire head to baffle detection. Only the opaque contents of the creature's stomach gave it away to Henry Moseley, and even these—mirabile dictu—seemed designed to resemble innocuous bits of floating seaweed.

Of course, camouflage must routinely fail or the ocean population would starve. Most deep-sea creatures have developed extraordinary visual powers to penetrate the myriad disguises of their wily prey. Bioluminescence has evolved independently more than forty times in the ocean, and eyes fifty times, suggesting a high-stakes arms race. Exceptions to this logic point to the humbling complexity of evolution. For example, how did *Challenger*'s prize lobster, the *Willemoesia*, come to be blind? Our answers are little more advanced than the German zoologist for whom it is named. With the enormous energy expenditure required for optical advantage in the deep sea, it's plausible *Willemoesia* gave up its struggle to defeat the genius of transparency the better to focus on perfecting its other senses. Cruising the sea bottom—a flat, populated plain more like an African savannah than the featureless miles of ocean above it—a lobster might rely on bumping into something edible. The ocean floor . . . where the invisible lead the blind.

Moseley's observations on the conspicuous transparency of pelagic marine life are the first in the scientific record. Arguably the *Challenger* philosophers' great achievement of their first year, however, was a related discovery, of which the spectacular light shows off the African coast were a nightly clue. Their around-the-clock observations, for the greater part of twelve months at sea, demonstrated beyond doubt that a grand odyssey of marine life (called diel vertical migration, DVM for short) did indeed occur on a daily basis, beginning at sundown when the sea surface rapidly filled with luminescent animals intent on feeding, and concluding with their reverse descent back to the cold depths at dawn. Creatures of all kinds—dinoflagellates, crustaceans, fish—participated in this epic commute.

The distance travelled? For tiny plankton, swimming 150 feet to the algae-rich surface and back is the equivalent of an average person hiking forty miles twice a day every day of their lives. On a sonar screen, it can look like the entire seafloor is rising up to the surface. Even in scientific parlance—not given to hyperbole—the DVM counts as "spectacular behavior."[10]

The trigger is light: with every dawn, all edible creatures at the sea's surface flee the revealing power of the sun's rays. The object of the never-ending traffic is to maintain a constant low-light intensity in the creature's environment best suited to their camouflage capabilities. The collective motive is likewise simple: to eat while avoiding being seen and eaten. We use "fishbowl" as a metaphor for an uncomfortable level of public visibility. Pity the legions of daily migrating plankton, shrimp, and fish in their Atlantic-sized fishbowl.[11]

10 Meester (2009) 651.

11 In our Anthropocene oceans, particularly near shore, light pollution offsets the nocturnal trigger for the vertical migration of billions of organisms—with as yet unknown consequences; see Davies et al. (2014).

At the end of August, *Challenger* had crossed the Equator for the first time in her journey, en route to Brazil. The modern Navy scorned superstition, so no one was rudely dunked or had their heads shaved, much to the men's disappointment. But the heavens had anticipated the moment for some time: the North Star daily grew fainter in the night sky, and southern constellations displaced the old. The novelty was enhanced by the sudden exchange of day and night at this latitude. Not the lingering, soft twilights of Edinburgh. Here darkness fell in a moment, like a curtain.

Warm, tropical rain pounded the deck; the ship was tossed by squalls. Taking daily measurements of water temperature at depth, the Challengers identified the precise course of the Equatorial Current that tugged them back toward to the Americas. They stopped at Fernando de Noronha to botanize—as Darwin had forty years before—but found it entirely denuded of its native forests. Adding insult to injury, the Brazilian officer in charge of the penal colony insisted nothing be removed from the island. "If we see a butterfly," asked Captain Nares, "may we catch it?" The commandant said he preferred they didn't.

They had their fill of butterflies soon enough. Coal supplies were near exhausted as they approached the coast at Bahia. Bobbing uselessly on a calm, hot morning in mid-September, *Challenger* was suddenly beset by a swarm of *Heliconius*. Thick as a blizzard, the butterflies swooped and quivered in the air before expiring in their thousands on the deck, even penetrating the wardroom. The ship's boys delighted in chasing them through the rigging, while Wyville Thomson was struck by the patterning on their wings, appearing uncannily like "withered leaves" of the forest from which they had come. Camouflage as a first law of survival demonstrated once more, on land as on sea.

A wealthy British expatriate named Wilson was pleased to host the *Challenger* scientists at his forest estate above Bahia.

From the comfort of the verandah, they breathed the perfumed air and interpreted the dance of fireflies in the garden as a visual echo of the stars above in their "crystal dome." The scientists had become accustomed to luminescent displays at sea as a first principle of that medium. Why, they now asked themselves, was the same phenomenon so rare on land?[12]

Dawn arrived with tropical abruptness. In an instant, the stars faded and the great trees emerged as a black mass against the brightening sky. This first hour of morning was the best for exploring while the air remained fresh and cool. They heard the scream of toucans long before they saw them at the very top of the green canopy. But their guns were useless at that towering distance.

Lower down, at eye level, the naturalists observed the jewelled colors of the toucan repeated, with variation, in the butterflies fluttering by "like a loosely-folded sheet of intensely blue tinsel," and in a strange harlequin bug—red, blue, and yellow—adhered to the tree branches. On this day, or a later time, Suhm snatched his prize green parrot from the jungle maze. The bird was destined to an unnatural life, typical of exotic pets. Cooped up in a ship's cabin, no longer able to blend with the forest in which it was born, it learned to mimic human speech during the long crossing to Cape Town. Later, in his last letter to his mother, written from the killing heat of the Pacific, Suhm would compare himself to a parrot transported from the tropics to a gloomy German forest, where "the bird dies!"[13]

The talent for camouflage among marine animals is well-known to generations of schoolchildren from field trips to public aquariums, which express our collective love of marine exotica. For most visitors, it is the closest encounter they

12 Thomson (1877) 2:117, 136.

13 Thomson (1877) 2:139–40; Willemoes-Suhm (1877) xi.

will ever have with ocean life—the majority of our planet's biomass.

The aquarium was a quintessential Victorian invention. Two decades prior to *Challenger*'s voyage, beneath the crumbling Cretaceous cliffs of Devon, Victorian evangelical naturalist Philip Henry Gosse prowled the shingle beaches for evidence of divine ingenuity. He had found it on his knees peering into the cold rock pools at low tide. In his best-selling 1853 memoir, *A Naturalist's Rambles on the Devonshire Coast*, Gosse described gazing a long hour into a tidal pool on Oddicombe Beach, as into a clear miniature sea. With an evangelist's fervor, he described its rim crowded with a "bushy fringe" of corals and fern-like seaweeds stretching to the surface. Colorful sea anemones that adhered to the rocky bottom looked like chocolate flowers streaked with scarlet. The strange brittle star crawled among them as if on hands and feet, while little blenny fishes darted from the shelter of the weed fronds. From the dioramic beauty of a Devon tidal pool, Gosse took the idea of the aquarium, which he popularized for Victorian households. With the proper balance of plant and animal life, ocean ecology could be simulated in a two-foot tank, at a reasonable price.[14]

The sea anemone (of the phylum Cnidaria), with its suggestion of brilliant floral arrangements, emerged as the unlikely star of Gosse's new drawing-room marine entertainment, crowded among the china mermaids and miniature castles of its little underwater grotto. In the 1850s and 1860s, dozens of suppliers sprang up to connect London customers with colorful, tentacular specimens scoured from beaches up and down the south coast. The self-educated Gosse contributed a definitive scientific assessment of the anemone that earned him a coveted place among the natural philosophers of the

14 Gosse (1853) 55. Credit for the original aquarium must go to Anna Thynne, who bred madrepores in a glass tank in her drawing room in the cloisters of Westminster Abbey in the late 1840s.

Royal Society, but he was at heart a popularizer. His insistence on observing living nature itself—not merely desiccated specimens in the laboratory—captured the public imagination and established a template for *Challenger*'s fieldwork practice. For the Victorian middle class, who required a wholesome recreation to occupy long holiday afternoons at the seaside, filling buckets with slimy anemones became a ruling passion. Gosse grew accustomed to receiving dripping packages in the mail—anemones sent by his fans.

Gosse's reputation as a pioneer of Victorian marine science suffered permanent eclipse as a result of his son Edmund's devastating memoir published in 1905. That grim oedipal portrait, however, is unrecognizable in his own autobiographical works from which Gosse emerges as a cheerful proselyte of the unregarded sea. The panoramic view from a Devon beach, he argues in his once-famous book, is more rewarding than any landscape because the ocean's tidal rhythms and ever-restless surface "are so much like the phenomena of life." At the other end of the scale, an hour spent in intimate study of a rock pool guaranteed "gratification of our sense of beauty." For Gosse, the sea engrossed the sublime and the beautiful in one shimmering element.[15]

But Gosse's beach philosophy did not extend to emerging theories of life. While his research with the microscope uncovered clear analogies of form and function between the sea anemone and local madreporian corals—specifically the stinging probes from which the Cnidaria phylum take their name—he refused to consider this proof of their common origin. Popular paeans to the beauty of marine life, in this sense, served as Gosse's lyric counterargument to the remorseless logic of natural selection, for which the ocean was increasingly the focus of debate. The fashion for Gosse-inspired aquariums faded through the 1860s,

15 Gosse (1853) 154, 354.

but Darwin mania was only just getting started, ensuring the Devonshire evangelist's long slide into obscurity.

The aquarium-clear waters off the Bahia beach were no less a theater of color than the forest above it. While a party of enslaved Africans loaded the *Challenger*'s hold with coal, the philosophers took the steam pinnace, with its low draft, to explore among the coral reefs almost at touching distance in the shimmering clear water. With their handheld nets, they collected several crates' worth of crabs, sponges, and brightly striped tropical fish. But it required wading in and duck diving to the sandy floor to capture the expedition's single representative of *Hippocampus*, the famously reclusive seahorse and aquarium staple.

Seahorses enjoy the aura of rarity but are in fact found everywhere: from Australia, to the coasts of Africa, and myriad reefs and mangrove shallows in between. For all this ubiquity, and its equine associations, a seahorse is a poor swimmer. It is therefore much attached to its home and to its mate, and escapes predators only through disguise. Camouflage strategies—vital for survival everywhere in the ocean—take a different form along the world's shores, with their colorful, garden-like structures and dense clumps of kelp, rock, and weed.

Instead of bioluminescence or cryptic coloration, coastal creatures such as *Hippocampus* rely on mimicry to elude predators. The seahorse changes color schemes to blend into the background, be it a bright yellow coral or the mottled root of a mangrove tree. Alternatively, it might sprout little tassels from its body in imitation of surrounding weeds. The Challengers no doubt came upon their prize by sheer luck, stoic and alone, its signature tail wrapped around a ligament of coral or perhaps a blade of seagrass to prevent being dragged about by the tide.

The *Challenger* naturalists left no description of their seahorse's pigmentation when captured. Inevitably, they misidentified their find, believing it to be a new species. *Hippocampus* taxonomy has always been a chaotic business because a known seahorse is so easily mistaken for something new: its chameleon colors match only the happenstance palette of corals or seaweed at the moment it was snatched.

The *Challenger* scientists were familiar with seahorses as a major attraction of the first public aquariums in Victorian Britain. Seahorse flesh was a delicacy well-known to visitors to the Mediterranean, but it wasn't until 1859 that the first live, captive seahorses made their perilous journey via train—in a glass bowl on a man's lap—from Portugal to the London Aquarium. Londoners flocked to watch the seahorses link tails in their signature maypole dance in a giant glass bath supplied by thousands of gallons of seawater barged up the Regent's Canal and piped in via a quarter-mile-long hose.

The seahorse certainly hates aquarium life, since it is notoriously antisocial and seeks out its own kind only to breed. The female transfers her eggs to the male for fertilization—a sexual congress unique in nature. The first observed breeding of live seahorses occurred in Manchester in 1873 only weeks before the *Challenger* collected its own Brazilian seahorse. Unfortunately, no attempt was made to keep *Hippocampus erectus* alive onboard. Within hours, *Challenger*'s lone seahorse had lost its candy stripe pigmentation entirely to drift pale and lifeless in a glass tube, where it remains today in the humid backroom repositories of London's Natural History Museum.

The interest of that historical specimen has only increased over time, however, as seahorse species worldwide are now officially threatened with extinction, the first regularly traded fish to earn that designation. As other charismatic animals have found, it does not pay to be attractive to humans. Perhaps hundreds of thousands of seahorses are annually plucked from the reefs of the Americas and Asia to populate home aquariums, mostly in Europe and North America. A principal source for the ornamental seahorse trade is Bahia, Brazil, where the Challengers happened upon theirs.

Even more devastating to *Hippocampus* than the aquarium trade, however, is its imagined medicinal properties. Dried seahorse, reduced to powder, is a staple of traditional Chinese medicine used to treat everything from asthma to ulcers to erectile dysfunction. Seahorses are not fished directly but collected as secondary bycatch in industrial-sized shrimp nets. For every pound of shrimp harvested in a fine mesh shrimp net, up to ten pounds of other sea life—sponges, starfish, and corals included—are discarded. Seahorses, worth their weight, are held over, but hard numbers on *Hippocampus* mortality are elusive. Conservative estimates suggest some 25 million seahorses a year are fished from the world's oceans to supply a voracious pharmaceutical trade. Shrimp fishermen throughout Asia, whose trawls wreak havoc on the seahorse's coral habitat, are now reporting a dramatic decline in numbers.

The spectacular camouflage abilities of the seahorse are in direct proportion to its lack of athleticism. Like so many of its cryptic counterparts in the open ocean, *Hippocampus*, to escape its predators, channeled its energies into mastering disguise rather than fleeing danger. But in 16 million years of existence, crowned by successful migration worldwide, the seahorse never reckoned on trawlers' nets rioting blindly across its scenic habitat, let alone the invention of the aquarium. Ironically, in the billion-dollar market for traditional Chinese medicines, a premium is placed on seahorses of an unnatural pale color, where rarity equals potency. But white is also the color of a dead seahorse, for whom the delicate, evolved protections of camouflage have failed. For a pale seahorse—ground into powder and bottled on a pharmacist's shelf—to provide maximum relief, it must first be drained of all oceanic meaning.

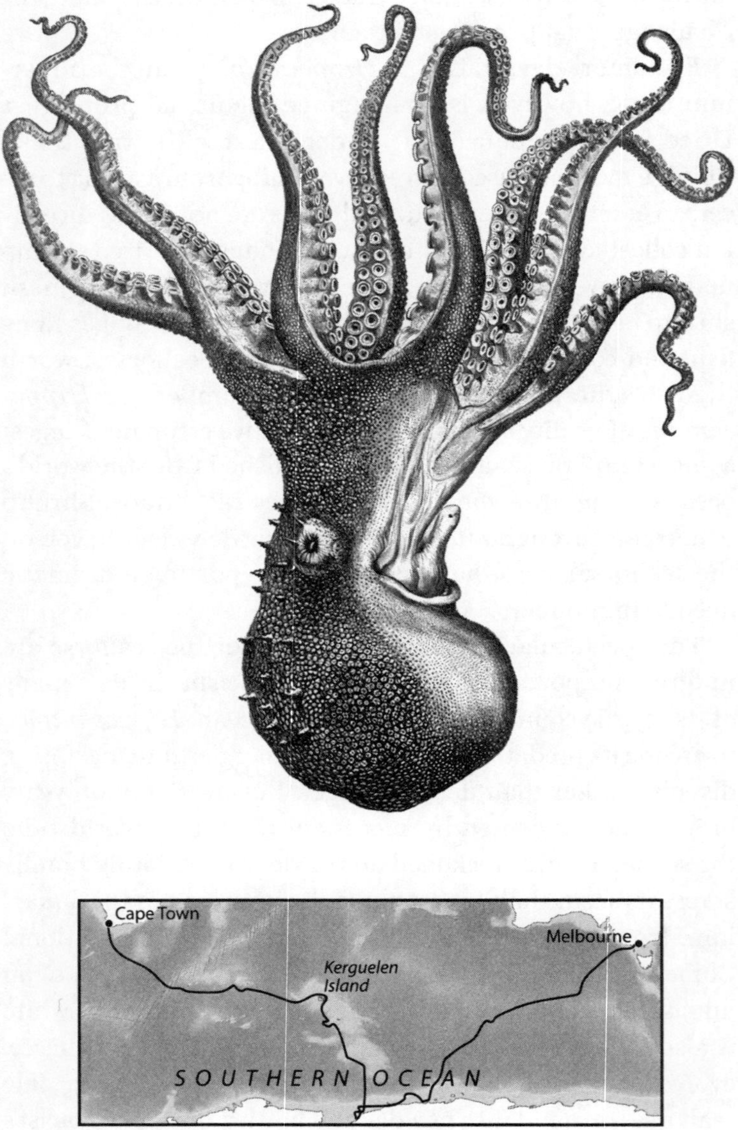

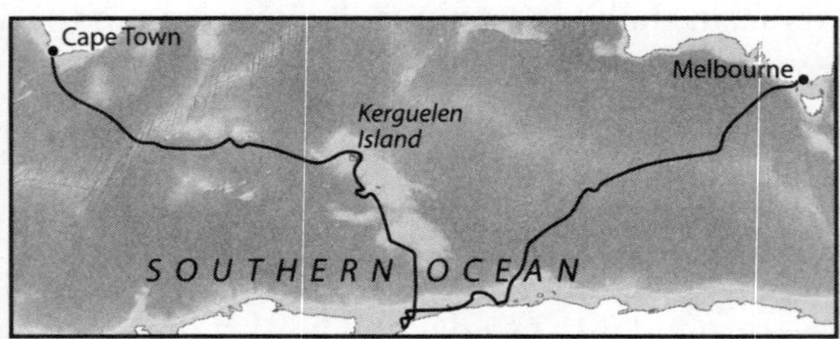

FIGURE 6.1 *Octopus pallidus*. From *Scientific Results: Zoology*, vol. 16, plate 1 (1886). (Inset) HMS *Challenger* Leg 6. Cape Town to Melbourne via the Southern Ocean and Antarctica. December 1873–March 1874.

CHAPTER 6

Heyday of the Octopus

Heard Island, Antarctica
53° S, 73°30′ E

The scientific personnel aboard HMS *Challenger* reflected the priorities of Victorian ocean research. Marine zoology stood supreme as the glamour science of the age, with physical oceanography a distant second—a niche field. Of the six *Challenger* philosophers, Thomson, the director, was a naturalist, as were Moseley, Murray, and Willemoes-Suhm. Thomson loved sponges and sea stars. Moseley claimed corals, Murray the tiny plankton, and Willemoes-Suhm crustaceans. But each of them was also by necessity a generalist, latching onto any and all eye-catching creatures thrown up by the net and assisting with rapid dissection and note-taking. Deep-sea biology aboard *Challenger* was a team effort.

Ocean chemist John Buchanan, by contrast, worked apart and alone. The naturalists enjoyed the airy community of the workroom adjacent to the captain's cabin, with windows onto the bright sea and sky. Buchanan's makeshift laboratory, meanwhile, was crowded amidships by the mainmast, with little ventilation or natural light. During *Challenger*'s three and a half years at sea, he collected several thousand samples of ocean water to which he devoted countless hours boiling for its chemical residues, carbonic acid in particular. His workspace was daily filled with choking gases.

Buchanan was a tall, bony individual, and the chemistry lab was smaller than his cramped sleeping quarters a deck below. Some nights, working late, he napped uncomfortably on top of a storage locker, careful not to smack his head on the bookshelves above. When conducting experiments at his workbench, he folded his writing desk against the window (formerly a gun port) to clear room. Aboard *Challenger*, Buchanan was known to be aloof and difficult, a reputation difficult to separate (in the historian's mind) from his pinched recent working conditions. All the more hilarious, then, the scene at the Cape Town Ball when Buchanan condescended to dance a reel with his fellow Scots John Murray and sub-lieutenant Campbell, his long arms and legs "flying about" in frantic disregard for rhythm.[1]

The situation of *Challenger*'s solitary chemist was awkward in more ways than one. While anchored in False Bay, Buchanan spent hours squeezed at his desk concluding a paper for the Royal Society on ocean acidity. A complicated assembly of tubes and vessels for measuring the absorption of carbonic acid in seawater stood on the workbench behind him. His 1874 paper today reads as narrow in its focus, but between the lines larger controversies loomed: How did acidity in the oceans vary with latitude and with depth? And what part did acidity play—in concert with temperature and salinity—in the radical new theory of ocean circulation that imagined a perpetual migration of polar undercurrents toward the Equator and a reverse course of lighter, tropical surface waters to the poles?

Aboard *Challenger*, the mere mention of a general ocean circulation—"sustained by difference of temperature" and independent of prevailing winds—was fraught with professional risk. The belief among Victorian scientists in a calm, unchanging deep sea set at a worldwide constant temperature of 39°F seemed itself imperturbable. Acceptance of a new oceanographic paradigm would not come easily.[2]

1 Channer [undated] 22.
2 Carpenter (1871) 54.

The origins of circulation theory dated to the summer of 1869 with the cruise of HMS *Porcupine.* On their northern leg, Wyville Thomson and his Royal Society colleague William Carpenter had been stunned to discover—via temperature readings at depth—the existence of *two* distinct subsurface currents in the abyssal channel between Scotland and the windswept Faroe Islands. Depending on the precise position of the ship, the bottom water appeared either the product of a cold current issuing from the Arctic or an adjacent warmer stream from Atlantic waters to the southwest. Taking his cue from Alexander von Humboldt's little-regarded theory of polar convection, Carpenter surmised that temperature differences, though slight, were propelling massive quantities of water in countervailing directions.

The Gulf Stream—that great Atlantic torrent responsible for Europe's mild climate—was already legend to the Victorians. But the new *Porcupine* data suggested—to Carpenter, at least—that the physics of the Gulf Stream were not unique to the North Atlantic but rather part of a wider, global network of deep-sea currents. From this insight, Carpenter deduced his theory of general ocean circulation—or "submarine climates"—which he defined as the "interchange of equatorial and polar waters continually taking place in the great Oceanic basins."[3]

In the winter of 1870, Carpenter had been invited to give a public demonstration of his theory at the Royal Institution in London. At the lectern where celebrity chemists Humphrey Davy and Michael Faraday had dazzled audiences in decades past, Carpenter unveiled a square glass tank filled with water and fitted with a heat lamp at one end and a wedge of ice at the other. He then directed his assistant to pour a vial of blue-colored water onto the surface followed by an equal volume of red-stained water injected at the bottom of the tank.

3 Carpenter, Jeffreys, and Thomson (1869–1870) 454; Carpenter (1870) 490.

The water responded instantly to the difference in temperature at either side of the tank, whereupon the audience was treated to the unmistakable sight of an overturning current in action. The blue water, cooled by the ice, descended rapidly before seeping along the bottom to the warmer side. It then rose up in a curlicue motion to the surface, replacing the red current, which had just then vacated the heated side for the cooler surface opposite. Differences in temperature drove circular convection, Carpenter declared. What was happening before their eyes in his little tank was true of the world's great oceans—a single ocean, indeed, connected by vast overturning currents hundreds of miles wide.

Wyville Thomson publicly dissented from his friend's view in their coauthored *Porcupine* report. He endorsed instead the establishment opinion of geographers James Rennell and Matthew Maury and ice age theorist James Croll, who argued that ocean currents were the product *only* of prevailing winds, as transatlantic sailors had reckoned from time immemorial. Thomson agreed that *Challenger*'s temperature readings at depth—conducted every two hours, every day—was necessary to settle the question but gave little credence to what he politely called Carpenter's "magnificent generalization."[4]

Carpenter's defensive tone in the ensuing paper war suggests he was piqued by the skepticism shown to his theory. For vindication, he looked forward particularly to *Challenger*'s identification of an "immense motor of Polar Cold" in Antarctica, the perilous leg of the expedition on which she was now to embark. Differences in ocean temperature, he believed, closely affiliated with saline changes in density, engineered the vertical transfer of oceanic heat across hemispheres—what scientists now call thermohaline circulation. Antarctic ice—alternately freezing and melting—provided the most dramatic contrasts in local temperature and salinity of any ocean on

4 See Deacon (1971) 306–32.

Earth. Would the Southern Ocean—remote and mysterious—prove itself the origin of the world's great undersea currents? Even the engineer of global climate?[5]

It would be understandable if John Buchanan—preparing his Antarctic experiments with accustomed meticulousness in his cramped laboratory—experienced a certain loneliness in contemplating the great questions he had been sent to answer. To begin, he drew up two new logs: the first to measure the acidity of polar ocean waters and a second for recording changes in the color of the sea's surface. Crystal blue signified a pure, fresher current, while murky green meant waters dense with life—and acids. At the workbench, he assembled an apparatus for testing the salinity of sea ice and for producing brine. This meant, one day soon, an adventure onto the ice itself in an open boat, armed with a pickaxe and bucket.

Finally, he unrolled from its tubing his oceanographic masterpiece-in-progress: a salinity map of the world's oceans. Changes in ocean chemistry followed the prevailing trade winds in their headlong escape from cooler subtropical zones to the Equator. Dense, evaporated waters, driven to the western Atlantic on either side of the Equator, showed bright red on his half-finished map, like seas within seas. If these great warm pools were destined to sink and help power ocean circulation, how much truer it must be for the rich salty waters of the Antarctic, theoretically the densest on Earth?[6]

Then there was the forbidding ice pack itself and the unexplored frozen continent beyond. Among all her fanciful divagations on the high seas, *Challenger*'s mission to the Antarctic held greatest interest for the British Admiralty. The far

5 Carpenter (1871) 68.

6 It is because of Buchanan's map that researchers have been able to identify the unusual rate of change in ocean salinity since the 1950s. An intensification of the hydrological cycle due to global warming has made salty areas saltier and fresh waters fresher compared to *Challenger*'s 1870s baseline; see Gould and Cunningham (2021).

Southern Ocean had not been visited since James Clark Ross's legendary voyage in the *Erebus* and *Terror* in the early 1840s, with Darwin's future champion Joseph Hooker aboard. Decades later, John Murray would draw on *Challenger*'s legacy to argue for a renewed push to the South Pole, heralding the age of Scott and Shackleton. But for now, Buchanan's interest in *Challenger*'s looming polar mission was oceanographic not geographic. Would chunks of ice, hacked from some passing floe in the frozen zone, offer up the secrets of their formation and prove the reality of a global ocean current system?

False Bay, true to its name, had offered a confusion of currents and a traditional argument for their wind generation. Wind out of the northwest supplied the bay with an influx from the cold Atlantic, while a southerly change, contrary to the logic of latitude, brought dense, warm water from the adjacent Indian Ocean. A sudden change in wind direction was capable of raising the temperature of the bay up to 6°C (11°F) in as many hours. Beyond the calm of the bay, however, even greater forces were at play.

Buchanan looked out across *Challenger*'s stern one morning a week before Christmas, 1873, and saw only open water. Africa had sunk from view in the course of the night. Three days out of Cape Town, the westward Agulhas Current with its monsoonal warmth suddenly switched direction, while the sea surface temperature plummeted. Peacoats, mittens, and thick wool drawers were ordered up from the hold. The Antarctic Circumpolar Current, swift and massive, had picked up *Challenger* and was urging her, bobbing and rolling like a bath toy, toward the frozen limits of the world. Their next safe port—Melbourne, Australia—lay 6,500 miles to the east. Most on board were quietly apprehensive beneath the usual bravado. No one was surprised when Captain Nares sent his ten-year-old son home prior to their Cape Town departure.

Heard Island gave them their first view of Antarctic ice, bleak and magnificent. Mist shrouded the mountainous summits of the island, so it was impossible to pinpoint the origin of the glaciers that overhung the black, volcanic cliffs. According to Buchanan's calculations on glacial melting, 1 percent of the great river of ice took the form of a bottom liquid brine, allowing the glacier to literally "flow," in slow motion, down the mountainside into the sea. Waves pounded the glacier's edge, sending huge icy shards thundering into the surf below.

Braving gale conditions, Buchanan and Moseley landed a boat on the beach to better study the great wall of ice. The black, windblown sand had scored intricate patterns on its surface and carved nearby rocks into the shape of trees. They covered their faces with scarves against the stinging sand. Captain James Cook had named the place Corinthian Bay, but the resident sealers—who met them warily, brandishing rifles—called it Whiskey Bay in tribute to their longed-for supply ship. These frontier hunters, whose sole job it was to drive unsuspecting sea elephants to the killing fields on the north side of the island, lived year-round in pits dug on the beach to protect them from the relentless westerlies. "I guess you are out of your reckoning," their chief said to them, bemused by a naval ship's presence at miserable Heard Island.

At 60 degrees south, *Challenger* encountered her first iceberg, which brought all but the most jaded hands on deck to bear witness. But even the polar novices soon grew nonchalant. Day after day through the end of February, ice-loving albatrosses glided above them, while the horizon was studded with icebergs of all shapes and size. The crystal atmosphere, free of dust and vapor, made distant objects appear vividly close. With his telescope, Buchanan observed the icebergs through the long polar days and lingering twilights. His reward was an ever-modulating palette of colors impossible in temperate climates, from the icebergs' first silvery, delicate unveiling at dawn to the gaudy reds and golds of sunset. The lashing sea had worn caves into the older bergs through which the sun reflected a dazzling azure blue. When a cloud passed over, the same ice retreated instantly behind an inky shroud, as if a painter had experienced a sudden change of mind.

The biblical storm of February 24 announced itself with a perfect rose sunset. Then the sky turned black and the sea an eerie color of steel. In blinding snow, surrounded by icebergs, *Challenger* labored helplessly for hours in the angry sea. Steepling waves, heaving up and down, gave the convincing illusion of driving them headlong to destruction. Next morning, while clearing the deck of wreckage, they passed an iceberg with a deep gouge in the shape of *Challenger*'s bowsprit—memorial to a frightening near-fatal collision.

Work aboard continued uninterrupted, though the temperature in the unheated workroom stood well below freezing. On the dredging platform, examining the day's haul, their faces burned from the driving snow crystals "as if they had been red hot." The ice pack—an otherworldly plain of whiteness stretching across the southern horizon—now presented itself. Bizarre sea temperature readings showed a warm current hundreds of fathoms beneath the surface, which explained both the termination of the ice pack at this precise latitude and the fate of the iceberg armada that surrounded them, imperceptibly shrinking as it drifted northwards beyond the Antarctic Circle. Buchanan, cocooned in every layer of clothing in his possession, rowed a small boat to the pack's edge. He spent only enough time on its treacherous surface to collect several melon-sized blocks before hurrying back on board.[7]

Here at the pack's edge, the acidity of the ocean, which had been falling since their departure from the Cape, spiked suddenly. Buchanan's reading—at this southernmost station—was its highest for *Challenger*'s entire voyage. With this "marked excess" of acidic waters, he presumed, must come greater salinity.

7 Thomson (1874a) 143.

He placed the sea ice in a porcelain dish and recorded the melt temperature as −1°C, well below that of fresh water. To test the salt content of the ice block appeared to be, at first glance, a simple task. In the end, it required more than ten years of follow-up experiments for Buchanan to conclude that the salinity of sea ice did *not* match the ocean water out of which it was formed. In fact, the process of freezing separated the fresh, light ice from an oozing, briny "mother-liquor," as he called it.

Winter sea ice formation, Buchanan reasoned, had the same concentrating effect on ocean salinity as the work of evaporation in the tropics. Subsequently, during the polar summer, the ocean heat absorbed in melting gigatons of ice created a cold, fresher "wedge" of water in the *Challenger*'s present path. This cold meltwater mixed with the saline waters of the surrounding ocean surface was currently sinking, all the while "diffusing in its descent the temperature of its formation." The interplay of these millionfold plumes of descending cold water with the warmer circumpolar current beneath produced a convective energy of globe-circling power. It mirrored exactly the process William Carpenter had demonstrated in his homemade tank before a fascinated London audience four years earlier.[8]

For Carpenter, ocean temperature was the all in all. Thermal variation provided the motor power for deep-sea currents and determined the distribution of animal life across its vast domain. Without the motive power of temperature difference, our oceans would be unrecognizable—an inert, tepid mass, mostly devoid of life. There would certainly be no penguins, fish, or great whales, which require a dynamic ocean with cold, nutrient-rich depths. Despite thousands of data points bearing out Carpenter's theories, Buchanan's post-*Challenger* publications hedge on the issue of a general oceanic circulation, perhaps out of respect for his chief who had been so public in his skepticism. In his official report on

8 Buchanan (1913) 25; (1874) 126.

Challenger's ocean water archive, drawn from all latitudes, Buchanan "deferred" on what bearing his density readings might have on ocean physics.[9]

Championing Carpenter's theory was thus left to the least heralded member of *Challenger*'s scientific staff—illustrator and naturalist-without-portfolio Jean Jacques Wild. Wyville Thomson had first crossed paths with the obscure Swiss schoolteacher in Belfast, where Thomson was professor of natural history. Wild's attractive illustrations for Thomson's *The Depths of the Sea* encouraged Thomson to offer him a berth on *Challenger* as ship's artist and personal secretary. Wild clearly saw the expedition as a career-making coup. From the outset, he looked for opportunities to publish. The ship was swarming with prolific, highly motivated biologists, so he looked instead to the new field of physical oceanography. He took time from his secretarial duties to pore over John Buchanan's logs of ocean temperatures, and presumably to discuss the problem of deep-sea currents with the tetchy Scottish chemist.

The result was Wild's remarkable *Thalassa: An Essay on the Depth, Temperature, and Currents of the Ocean*, published in 1877. The volume—in which he anglicizes his name as John James Wild—offers a lucid summary of *Challenger*'s ocean temperature data but also, more significantly, describes global undersea currents unequivocally in the language of Carpenter's vertical oceanic circulation. Currents are, Wild correctly argues, a product of "the difference of temperature, specific gravity and chemical composition of the water . . . all acting and reacting upon each other." In a direct riposte to Carpenter's nemesis James Croll, Wild holds up *Challenger*'s deep-sea temperature data as proof that "even in the absence of wind, the thermal circulation of the ocean would resolve itself into a system of surface and under-currents."[10]

9 Buchanan (1884) 12.
10 Wild (1877) 9, 53.

Of the *Challenger*'s fifty published volumes, forty are devoted, in part or full, to marine biology, while only two contain essays related to ocean physics and chemistry. Wild's *Thalassa* thus stands as both a rogue scientific publication—outside the *Challenger*'s fifty-volume canon—and its fullest, most forward-looking contribution to modern physical oceanography. In publishing it, the uncredentialled intellectual refugee from Zurich had taken professional risks unpalatable to his establishment colleagues.[11]

Thomson's opinion of *Thalassa* is unknown. Whether Wild alienated his patron in endorsing a heterodox theory of ocean currents, or suffered professionally more from Thomson's sudden decline and death in 1882, is difficult to judge. Either way, he failed to find an academic position on *Challenger*'s return to Britain, was reduced to the life of an itinerant lecturer in science, and died bankrupt in Melbourne in 1900, aged seventy-two.

In *Thalassa*, Jean Jacques Wild vividly describes *Challenger*'s epic journey across the Southern Ocean to his future adopted home via a vast "South Australian Current"—now known as the Antarctic Circumpolar Current. This mighty eastward-flowing, submarine river connects the Antarctic basin to the land-rimmed Indian Ocean from which they had come. Only at the southernmost reaches of their journey, at the ice pack, did they not feel its influence.

John Buchanan's color log—initiated for this Antarctic leg of their journey—was filled with descriptions of the turbid maelstrom of their polar passage and its unique hospitality to life. Carbonic acid levels rose proportionally as the sea surface temperature fell, while the translucent indigo waters out of

11 John Murray excepted. In an 1877 public lecture, Murray refers specifically to a "general ocean circulation" and to the expedition's finding, based on soundings in the Southern Ocean, that from there "the water is drawn northward over the ocean's floor to make good the deficiencies caused in the tropics by surface currents and evaporation." (1877) 352.

Cape Town turned to green almost overnight—a measure of the rich life beneath. To the surprise of all on board, the Antarctic open sea—not the fishing grounds of the Atlantic and Pacific coasts—proved itself the scene of *Challenger*'s richest hauls.

Abundant nature, in all her colors, ruled the Southern Ocean waves. Algae, mixed with floating excrement, surrounded the ship like an undulating green carpet in all directions. Microscopic diatoms—great yellow masses of them—poured from tow nets brought on deck. Tiny crabs, worms, and copepods crawled from the slimy ooze—the stench enough to make the hardiest philosopher retch. At other times, bright red plankton filled the nets, bursting from the strain of being hauled aboard. And around the ship, at all hours, rotated a noisy parade of predators—penguins, seals, and whales—insatiable gourmands of this movable feast.

During their three-month Antarctic crossing, *Challenger*'s dredge and nets disgorged fifty never-before-seen species of marine animals. Even more remarkable to the naturalists, however, was the proportion of polar creatures already well-known to them from high northern latitudes—the North, Baltic, and Arctic seas—and even the tropics. Deep-sea life was cosmopolitan, linked by Carpenter's worldwide system of circulating currents. For the long weeks they spent cruising atop the great eastward drift from Antarctica to Australia, the contents of the dredge became almost monotonously familiar.

Antarctic life held a further odd distinction, however: gigantic size. Enormous sea spiders, shrimps, and crabs of all kinds issued from the dredge. Were these living relics of a distant epoch, Suhm wondered, when giants ruled the seas? *Challenger*'s collections from the world's coastal shallows—from Portugal, to Brazil, to South Africa—seemed modern and diminutive by comparison to these brawny polar survivors.[12]

12 The inverse is actually true. Most deep-sea creatures are descended from shallow-water species, a conclusion intuited by John Murray. He pointed to *Chal-*

Their biggest Antarctic prize was an octopus fished from the depths west of the Crozet Islands. It had a rose-colored head and ghostly, web-like membranes between its massive arms. A creature of true alien rarity—the size of a man rebuilt as a pink umbrella. Suhm remembered the octopod's design from an illustration he had once seen in a Danish journal, and so consulted the onboard library. Sure enough: a solitary, diminutive counterpart to this giant *Cirroteuthis* (*Inopinoteuthis*) had been fished off Greenland in the 1840s. The North and South Poles were in communication via this beautiful monster, but across what precise axis of space and time he didn't dare to guess.

Aristotle was the original beach philosopher—and the first to leave a written description of the octopus's habits. On the island of Lesbos—along a placid, teeming lagoon of the Aegean Sea—the great polymath combed the beaches for anemones, sea cucumbers, and crabs. He quizzed local fishermen and watched sponge divers—ancestors of today's scuba scientists—plunder the reef thirty feet down. They told him—among innumerable tidbits of local wisdom—that ubiquitous starfish were a menace to their operations. The lucrative scallop population, he also learned, had been overfished to extinction, the first recorded human-driven crisis in marine life.

In two years' research on Lesbos, Aristotle created a written inventory of more than 180 marine species, mostly fishes, starfish, sea urchins, as well as mollusks of all varieties. More than that, he gave structure to his curiosity, crafting detailed introductory lectures for his students back in Athens. Marine animals were complex beings, not merely food and sponges for household consumption. Their color and variety were astounding, and their anatomies, however bizarre, suggested

lenger's "considerable evidence for supposing that migrations have taken place from the shallow waters of the continents into the deep waters of the ocean basins, and that such migrations have been going on for a vast time." (1877) 133.

mysterious connections with other forms of life. Most importantly, their body structure, behavior, and habitats were all proper objects of philosophical inquiry.

Aristotle's taxonomy of marine life, from his epic *Historia Animalium*, contained no higher orders, simply *genera*, of which his favorite were the agile, soft-bodied cephalopods: cuttlefish, squid, and octopus. The location of their heads, squeezed between their "feet" and bellies, fascinated him, as did analogies between their body parts and other creatures he knew. The cuttlefish's esophagus was "like a bird's." The stomach, with its twisted shape, resembled spiral shells he found washed up on the beach. Beneath the stomach were white protuberances "like breasts," while sex was conducted via a tangled union of mouth and tentacles. The resultant eggs looked like "big black myrtle-berries" (or perhaps "grapes"). Most of all, Aristotle admired the cephalopod for its cunning: confronted by a predator, it changed color or shot its black ink to the rear before retreating into the dark cloud, like a magician's disappearing trick.[13]

Octopods and squids—from the class Cephalopoda—occupy a vital trophic role in the world's oceans, as both predator and prey. Omnivores with short life cycles, their numbers are prone to fluctuation. Crossing from the Southern Ocean to the far western reaches of the Pacific in early 1874, the Challengers encountered a dazzling variety of cephalopod adaptations, many of them unique to Antarctic waters. The very day following their discovery of the octopus *Inopinoteuthis magna*, a giant squid tumbled from the dredge looking much like a giant vegetable: atop a cylindrical body with bulging eyes, its eight arms sprouted like cabbage leaves. Then, off Heard Island, where "the bottom of the sea seemed to be teeming with

13 Aristotle (1984) 832, 838, 855, 868, 968. Centuries later, in 1831, Charles Darwin was equally impressed by the abilities of an octopus he collected off the Cape Verde islands, one of his first prizes of the *Beagle* voyage.

animal life," emerged a family of new octopods, two adults, two juveniles, with parachute webbing and globular heads.[14]

By the first week of March 1874, they had rejoined the warm current pulsing east toward Tasmania. The tow net, dragging the surface waters at *Challenger*'s stern, scooped up a bathful of planktonic slime in which they discovered a new squid, named for Suhm, who dissected and drew it. More new octopods followed: from the deep-sea south of Australia to the coastal island waters of Bass Strait to Twofold Bay, near Sydney. There they came within view of a derelict whaling station—a reminder that their Southern Ocean tour intersected with both the migratory pathway of the great whales and the ships that hunted them. The food web of the Victorian Antarctic, where sperm whales dined on hundreds of tons of cephalopods daily, included humans as its top predator.

The Pacific Ocean whale fishery might be more famous, but whalers and sealers had already decimated the mammalian stocks of the Southern Ocean by the time of *Challenger*'s visit in the mid-1870s. Prehistoric hunters had long ago driven coastal seal populations to their island sanctuaries in the far south, while the warm tongue of the Indian Ocean current penetrating the Antarctic Circle lured sperm whales into polar waters. Operating out of Madagascar, American hunters massacred seal populations on the Crozet and Kerguelen islands to extinction, delivering thousands of barrels of whale oil annually to European markets. British whalers in turn plied the sheltered bays of southeast Australia. Tasmanian coastal waters teemed with right whales at the time of Europeans' first arrival around 1800. In Hobart, on the River Derwent, venturing out in a small boat had been considered dangerous. A short half century later, the whales were gone.

During his Southern Ocean crossings in the early 1840s, James Clark Ross in HMS *Erebus* had been routinely mobbed

14 Murray (1895) 1:487.

by whales, sometimes buffeting the hull. Putting Ross's account alongside *Challenger*'s a generation later, one is struck by the relative scarcity of whales encountered by Thomson's team (and a corresponding abundance of cephalopods). Henry Moseley's observation of two giant blue whales cavorting near the ice pack was sufficiently unique to warrant mention in his diary. Their bodies were so enormous that he could see the reflected sunlight from their sides illuminating the glass-green waters around them.

A plausible scenario suggests itself: the decades-long slaughter of toothed whales and sea elephants in the Southern Ocean had released the cephalopods from their principal predatory threat, allowing a spike in numbers. Jules Verne had already predicted as much. French naturalist Alphonse Toussenel, featured in *Twenty Thousand Leagues Under the Sea*, describes an inevitable "plague" of cephalopods and jellyfish once the oceans are cleared of whales.[15]

Today, octopus and squid are indeed flourishing, though not due to whaling, which is largely banned. Instead, overfishing has emptied the global oceans of their midsize rivals—tuna, grouper, cod, etc.—thereby opening a banquet of possibilities for cephalopod expansion. The impacts on human diet are significant. A simple glance at a seafood menu will reveal that, with 90 percent of the world's fisheries at peak or exhausted, cephalopods represent an ever-increasing fraction of the global catch.

It's a new chapter in an old success story. Opportunists par excellence, cephalopods have flourished in the world's oceans since the Cambrian period, predating the arrival of seals and whales by hundreds of millions of years. Signs are that they'll outlast the newcomers yet. Experiments have shown octopods acclimate better than ocean mammals and fish to the increasingly acidic waters predicted for the Anthropocene.

15 Verne (1876) 281.

Warmer ocean temperatures likewise favor the cephalopods by accelerating their life cycles and inflating body size. When the heat becomes unbearable, squid larvae will simply drift with the currents to populate cooler waters. In a worse crisis—be it hypoxia, plastic pollution, or deep-sea mining—the cephalopod is able to shrink its body size, reduce numbers, or even starve itself, and then magically bounce back when conditions improve. With rapid population turnover, languid mobility, and a scavenger's diet, cephalopods will outstrip many of their longer-lived, slower-growing vertebrate rivals in adapting to ocean catastrophes of all kinds.

Among the so-called weedy species of our Anthropocene seas, destructive blooms of jellyfish have attracted more alarm than cephalopods, which are, at least, better menu items. Jellyfish, recent research suggests, might have been the first multicellular life forms, narrowly beating out the sponges. Whatever the case, the uncanny similarity of ancient and modern jellies points to long-term evolutionary success. Jellyfish are to be found everywhere in the oceans, at all depths. The Challengers collected several deep-sea jellyfish including, off the coast of New Zealand, the spectacular crown medusa *Periphylla*, which Ernst Haeckel later rendered in gorgeous technicolor and with thirty pages of loving description. But the *Periphylla* have now penetrated Norwegian fjords in plague numbers, and their gooey brethren are clogging fishing nets in the Sea of Japan.

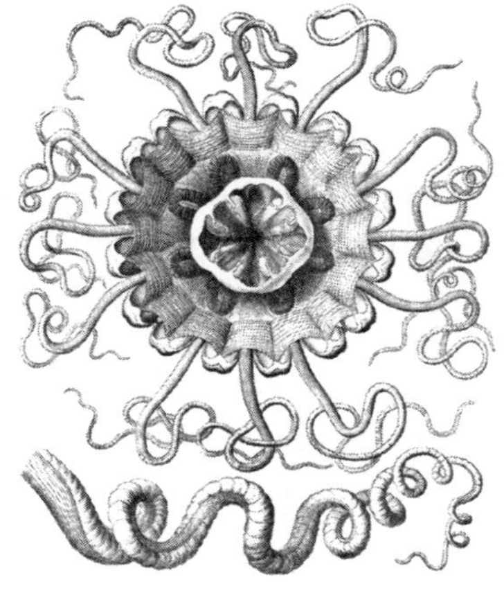

FIGURE 6.2 Ernst Haeckel's illustration of *Periphylla periphylla*. From *Scientific Results: Zoology*, vol. 4, plate 19 (1882).

Like the cephalopods, jellyfish thrive under selection pressure and have benefited from the human appetite for fish flesh. Jellies are currently colonizing new niches cleared of their rivals by trawling nets the size of football fields, and they seem immune to higher temperatures. The decline of sea turtles has eliminated a main predator, while hypoxic "dead zones," which kill off most other forms of life, are party spaces to the jellyfish. It's almost certain, too, that human-engineered coastal sprawl—from concrete piers and floating docks to glass and plastic trash—has opened new breeding habitats for jellyfish polyps, which adhere to artificial hard surfaces by the billions.

Overfishing, in combination with climate change and human-built sprawl, may have epochal consequences for the ocean food web. Paleo-oceanographers have drawn a contrast between the tepid, "low energy" Cambrian seas—dominated by jellyfish and their tiny flagellate prey—and our diverse, high-energy oceans populated by whales, fish, and larger zooplankton, particularly diatoms. The primeval ocean of the Cambrian greenhouse bears little resemblance to today's. Giant supercontinents spanned the Equator with warm, shallow seas lapping the coasts. Much of marine life as we know it did not yet exist. Likewise, there was no Gulf Stream or global overturning circulation to energize ocean nutrients and disburse life. But the Anthropocene ocean is trending Cambrian. In our warming, nutrient-depleted seas, flagellate plankton may swim deeper to find food where free-floating diatoms cannot. Advantage jellyfish. With the base of the marine food web undermined, the balance might swing back toward the gelatinous seas of 500 million years ago.

One thing's for certain: a Cambrian HMS *Challenger* could not have sailed a brisk ten knots for thousands of miles east of Africa to the very strait where the Antarctic continent finally separated from its Australian neighbor 30 million years ago. That tectonic rupture unleashed the circumpolar current that created the Southern Ocean as we know it, and ultimately the modern global climate system of tropical, temperate, and polar zones.

As they drew nearer to the Australian coast, Moseley noticed the sea surface fauna around the ship changing. At night, their wake was lit by the glowing *Pyrosoma*. The floating kelp *Durvillea*, too, returned, sometimes steered by its dedicated parasite, the *Phronima*, which had established residence on the weed with its young. He observed the amphipod's tail working vigorously through the green water like a stern propeller. Trawling in the warmer Indian current produced an unusual prize: the so-called obese dragonfish *Opostomias micripnus* (actually quite slender) with phosphorescent organs for eyes, like headlights.

On the afternoon of March 15, 1874, the Challengers spotted their first ship in three months off the port bow. They had crossed from the wild, deserted Southern Ocean to the Pacific, optimistically named for its peaceful waters and synonymous with island allure. The frostbitten Challengers gratefully put away their thick peacoats and mittens. They looked forward to fine weather and the hospitality of coastal settlements they recognized as civilization.

Sailing into Port Phillip Bay, the frontier town of Melbourne appeared as a pencil-drawn outline on the horizon. Suhm noted "many large medusae floating past the ship"—but he paid them little mind. These ancient jellies—brainless, watery creatures—had adapted superbly to the modern marine ecosystem with its powerful currents, as they will surely revert again if warming melts the polar ice and slows the majestic, worldwide ocean circulation to a dribble. With no motive power of their own, the resurgent jellyfish will drift with any residual currents, or simply multiply where they are. Instead of fast-flowing waters "teeming with life," the view from the deck of a future *Challenger* vessel might well be a gelatinous goo—jellyfish from hull to horizon, floating listlessly in the tepid bathwaters of a neo-Cambrian sea.[16]

16 Murray (1895) 1:530.

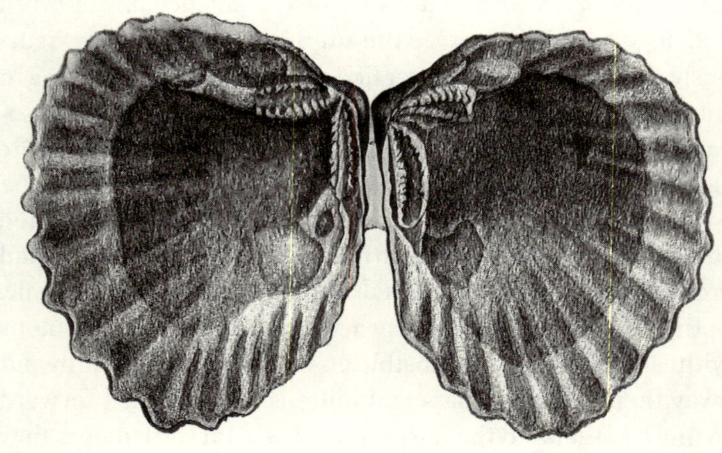

FIGURE 7.1 The holy grail of mollusks, *Trigonia* (*Neotrigonia margaritacea*). From *Voyage de la Corvette l'Astrolabe pendant les années 1826–29, Zoologie: Atlas* by Jules Sébastien César Dumont d'Urville, plate 78 (1833). (Inset) HMS *Challenger* Leg 7. Sydney to Cape York via the South Pacific. April–September 1874.

CHAPTER 7

Brooch Clams and Hairy Mussels

Hawkesbury River, New South Wales
33°33′ S, 151°18′ E

Challenger steamed into view of the scenic sheltered bays of Sydney Harbour on Easter Monday, 1874. At Farm Cove, reserved for naval vessels, the shore was filled with colorful picnickers, who soon swarmed aboard. Days later, the stream of nosy visitors full of questions had not abated. The philosophers, seeking relief, fanned out across the famous harbor with its countless creeks and coves overhung by eucalypts—a naturalist's paradise.

Suhm was the most desperate for isolation. Professor Thomson had relented on his German publication ban, leaving Suhm free to communicate his discoveries to the prestigious *Zeitschrift für Wissenschaftliche Zoologie* in Leipzig. He gathered up his papers and microscope and escaped the throng of Sydneysiders to Botany Bay, ten miles south, where he rented a cottage by the water.

The contrast with life on *Challenger* was dramatic—and restorative. Shipboard routine had exhausted him: late nights in the workroom notating the day's catch, only to be hounded awake before dawn by the clatter of the cleaning detail inches above his head. This relentless naval existence that murdered sleep, combined with his propensity for seasickness, meant he had barely known a day's proper health at sea since leaving Sheerness. The Antarctic cold had energized him, temporarily. But their coming transpacific odyssey—months in the steamy island tropics—would spell his undoing.

For now, however, the young German luxuriated in his wild retreat in a gum-tree forest by the water, surrounded by screeching cockatoos, butterflies the size of his palm, and giant fruit-eating bats the locals called flying fox. Rambling across the dunes, he nearly stepped on an echidna. He slept twelve hours a day on a great bed hung with mosquito curtains and went swimming in the fresh, cool bay. His Brazilian parrot—excited by the forest surrounds—welcomed his dripping return with a volley of expletives.

The landlord tried to lure him out to hunt kangaroos and opossums, and met his demurral with barely polite surprise. He could not comprehend the young foreigner's preference for books and quiet contemplation. In a letter to his mother, Suhm congratulated himself on having conquered for now "the furies of unsatisfied existence." But the solitude of empty days tended to blur the mental present with the past. At moments, his mind yielded to the sad thought that his Leipzig friends had forgotten him, while he brooded on *them* from his Australian idyll. In the letter, he described memories of his once-in-a-lifetime voyage like a pageant: a "chaos of people and things—a colorful abundance rising and falling in front of me."[1]

In Melbourne, he had been glad to meet with a German botanist, Baron Ferdinand von Müller, who showed him the zoological gardens, including the rare Tasmanian devil led about on a leash. The city fruit market was no less a zoo, with cockatoos and kangaroos for sale in cages alongside mounds of peaches, pineapples, and guavas, all at boomtown prices. Baron Müller's parting gift was a nautilus shell, fossilized, for the *Challenger*'s collection.

The Baron introduced him in turn to a Professor Frederick McCoy from the university, "a small reddish bon vivant," who had given up dull research in favor of burnishing his personal cabinet of curiosities. Suhm pondered a fossilized

1 Willemoes-Suhm (1877) 91.

FIGURE 7.2 An 1880s view of Sydney Harbour shows an already heavy industrial footprint and the dominance of steam shipping. Julian Ashton, *Circular Quay, Sydney* (1888).

sea lily ranged alongside stuffed Australian parrots and a trio of enormous gorillas. One exhibit he longed to see, but which was no longer in the professor's possession, was a single *Trigonia* clam—a living fossil of the Antipodes—recovered from a beach cliff in nearby Torquay. McCoy had kept the miraculous shell as long as he dared before forwarding it for display in London.[2]

While at anchor in Sydney Harbour, *Challenger* expended every spare moment, and precious steam power, in wholly unscientific hunting for *Trigonia* shells. All on board showed a sudden interest in the tedious work of dredging because, in colonial New South Wales, the ancient *Trigonia* commanded high prices as a fashion accessory. The women of Sydney wore their pink shells as earrings inlaid with gold or as "brooch clams." Officers and men alike prized them as gifts for their onshore sweethearts.[3]

Suhm requested the steam pinnace and set out alone in quest of *Trigonia*. The shiny white variety were common enough but less attractive. At the end of a long day's dredging along a treacherous reef near the heads—called the Sow and Pigs—he discovered a crop of rare beauties, colored a lustrous purple on their inside shell. He joked with his colleagues about his natural genius for *Trigonia* collecting, then placed them carefully in a box to be fashioned into a necklace for his mother.

The *Trigonia* was that rare natural object valued equally for its beauty and scientific importance. On Cook's Pacific voyages of the 1760s and 1770s, naturalists and seamen alike gathered tropical shells for sale to wealthy European collectors. But in 1799, the French taxonomist Jean-Baptiste Lamarck had declared an end to shell collecting as a mere "vain object of

2 Willemoes-Suhm (1877) 95.

3 Moseley (1892) 239.

amusement." Fossilized mollusks uncovered at sites from the suburbs of Paris to the Alps had disclosed a deep-time Earth history marked by successive inundations of the continents and the abrupt overturn of fauna. Seashells possessed geohistorical secrets.

Lamarck, however, like the majority of his contemporaries, resisted the idea of catastrophic extinction events as the organizing principle of Earth's history. To effectively defend against the extinction paradigm, he impressed how "very essential" it was "to seek out and identify living analogues to the great number of fossilized shells" now unearthed. The *Trigonia* shell was among the most cosmopolitan and diverse in European strata, abundant throughout the Mesozoic but with no apparent descendants beyond the abrupt event that doomed the dinosaurs. To find a living *Trigonia* clam would sink extinction, with its uncomfortable religious and political overtones, and affirm Lamarck's transformist view of the infinite adaptability of species in response to environmental stress.[4]

Three short years after Lamarck's essay, a young naturalist named François Péron—travelling with a French expedition—discovered a separated *Trigonia*-like shell on a beach at Bruny Island in southeast Tasmania. The iridescent shell bore distinctive ornamentation familiar from the fossil record: curved ridge-lines diverging from a central rib plus an intricate hinge. Back on board, Péron presented his discovery to his senior colleague, René Maugé, who was confined to his cot with dysentery. At the sight of the living *Trigonia* shell, even a half of it, the dying man burst into tears. Lamarck rushed into print to announce the existence of *Trigonia margaritacea*, which he named for its pearly luster. "Among the numerous conquests naturalists are making, on a daily basis, among the productions

4 Lamarck (1799) 63. *Trigonia* as a genus is permanently extinct. The extant genus, identified by the Victorians in Australian waters, was reclassified as *Neotrigonia* in 1912.

of nature," he wrote, "none is more interesting than one we held out so little hope for."[5]

With proof that the *Trigonia*, having disappeared from northern oceans some 65 million years ago, was alive and well in the southwest Pacific, it remained only to find a complete, healthy specimen. Enter Jean Quoy and Paul Gaimard, naturalists aboard the French corvette *L'Astrolabe*, who in late 1826 anticipated *Challenger* by fifty years in trawling Bass Strait and Westernport Bay. They searched day after day across "une vaste espace" for the fabled mollusk without success. Just when hope seemed lost, on a calm night off Cape Dromedary along the New South Wales coast, the net pulled a fresh trove of clams onto the *Astrolabe*'s deck. Barely distinguishable by lantern light among the heap was a single living *Trigonia* with its distinctive pink-purple luster on the interior of the shell.[6]

A last twist of the tale lay in store for the first modern *Trigonia*. *L'Astrolabe*, with veteran Pacific navigator Dumont D'Urville at the helm, made heavy weather of her passage from the Bay of Islands in New Zealand east to Tongatabu—the same course followed a half century later by HMS *Challenger*. Armed with the charts of Cook and d'Entrecasteaux, D'Urville risked the patchwork of harbor reefs to anchor at Nuku'alofa. But the *Astrolabe* was quickly stranded by the ebbing tide. On the third day, with the ship's masts tilting toward horizontal, D'Urville passed the word for an orderly abandonment. This included nailing up the ship's weapons in the hold to prevent his men seeking new employment as armed militia for rival Tongan chiefs, who observed the impending French disaster from their canoes.

The ship's company rushed to the boats with no room permitted for possessions. This order extended to Quoy and Gaimard, who had a year's worth of meticulously ordered spec-

5 Péron (1807) 1:240; Lamarck (1804) 551.

6 D'Urville (1830–1834) 3:475.

imen jars crammed in their cabin. After a quick conferral, the two naturalists agreed to comply with D'Urville's order—with a single exception. Secreted in his pocket, Jean Quoy was determined to salvage *Trigonia*, the expedition's great scientific prize, destined to enter nineteenth-century scientific lore as a contested, protean symbol of the history of life.

Quoy's hard-earned *Trigonia* proved compelling to a generation of European naturalists divided over Lamarck's transformist theories. Louis Agassiz opened his 1840 monograph on fossil mollusks with an essay devoted to "les trigonies." Opposing Lamarck, Agassiz interpreted the discovery of the living fossil in the Tasman Sea as proof of catastrophic overturn in biological history, not the continuity of species. Because no *Trigonia* shell had yet been identified from the Tertiary period—*between* the late Cretaceous and the modern *Trigonia magaritacea*—it was not possible in 1840, Agassiz argued, "to uphold the principle of the filiation of species of the same type across different geological epochs." [7]

It was a preemptive strike against Darwinism. For Agassiz, who published his catastrophist theory of ice ages that same year, the patchiness of the fossil record represented not gaps in modern knowledge but rather "shocking interruptions" in the history of life itself by which different regimes of plants and animals successively inhabited the Earth. The modern *Trigonia* clam marked an instance of a species' "surprising return" after prior extinction. Hedging his bets, Agassiz denied all proof of evolution even *were* a Tertiary period *Trigonia* to be found. The grand mosaic patterns of life could not be the product merely of incremental change over time; rather, each epoch formed an "organic whole" spontaneously brought into being at its zero hour.

Tertiary *Trigonia* deposits were soon found to fill the embarrassing gap, most notably from the famous beach cliff in

7 Agassiz (1840–1845) 1:3.

Torquay. Possibly because its first custodian, Frederick McCoy, was a fierce opponent of the just-released *Origin of Species*—or more likely because his preference was, as Suhm noted, for "undisturbed relaxation"—the startling discovery went unpublished.

The Torquay *Trigonia* eventually made its public debut in Melbourne in 1861. From there it travelled to the London Exhibition of 1862, where it came to the notice of a progressive geologist named H. M. Jenkins, who duly publicized the Miocene *Trigonia subundulata* as fresh evidence of Darwin's theory of "descent with modification." Jenkins concludes his brief essay with reflections, common at the time, on the apparent asynchrony of northern and southern zoology—how the Jurassic period in Europe lived on in the marsupials and mussels of Australia and New Zealand.

And with that, the *Trigonia*'s public career as a "living fossil," fetish object, and scientific debate point, came to an abrupt close. HMS *Challenger*, a trailblazer in so many fields, had arrived altogether late on the trigonian scene.

Bivalve mollusks like *Trigonia* are ecosystem engineers reliant on a brackish, estuarine mix of fresh and saltwater. As filter feeders, oysters, clams, and mussels are critical to water purity. As reef builders, they provide a miles-long, three-dimensional habitat for inshore fish, sponges, and a myriad invertebrates. Their loss is invariably devastating to the biodiversity of any coast and to its protection from the relentless weathering of storms and tides.

Like Willemoes-Suhm, Henry Moseley felt an invitation to explore Sydney's watery byways for examples of its famous oyster beds. He rode one morning along a high, sandstone plateau studded with homesteads of the colonial elite before coming to a narrow chasm. He clambered down through a forest of luxuriant ferns, lyrebirds in a bell-like chorus around him. At the bottom, he emerged from a shaded gully onto the slippery banks of Berowra Creek, a broad, meandering tributary of the Hawkesbury River.

Here, twenty miles from the Pacific Ocean, marine life abounded. On a shallow sand bank, he observed stingrays basking in the sun. Nearby, a shoal of saltwater mullet sought refuge from a pursuing dolphin. He caught several of the grey fish with a single dip of his net. Then a swarm of jellyfish drifted by. One might, he wrote in his diary, "sit on a gum tree and fish for sharks." In this magical spot, it was impossible to tell where the sea ended and the river began.[8]

Land and water, too, were in intimate relation at Berowra Creek. The sandy bottom was crowded with rotting tree branches and leaves. The birds swooping about his head might find burial there as would, from time to time, an overambitious wallaby. Not to mention the great oyster beds, by turns basking in the shimmering air then immersed in the advancing tidal waters. A future geologist, he reckoned, could make their career in such a place, where the bones of marsupials, birds, fish, and cetaceans—together with flowers and shells and insect impressions of all wild kinds—would find permanent rest arranged side-by-side in an earthen gallery for the ages.

The mollusks of Berowra Creek, in particular, caught Moseley's attention. The rocks studding the creek bank were smothered with clumps of oysters, while large heaps of discarded shells lined the banks. At every juncture of a little creek or inlet with the main river loomed a vast shell midden—the legacy of generations of meals consumed by the Gadigal people, who had inhabited this place seasonally for millennia but had recently been uprooted by European invaders.

For Moseley, who had ambitions to publish a popular account of his naturalist's travels in the style of Darwin's *Beagle* memoir, the middens were an exciting find. Darwin's friend John Lubbock (an archaeologist who invented the categories

8 Moseley (1892) 235.

Neolithic and Paleolithic) had recently published on the shell middens of ancient peoples of the Baltic coast. Here, before Moseley's very eyes, was the work of their Australian counterparts—the shellfish architects of the Antipodes.

In a cave above the creek, Moseley found more shell middens, but also wall paintings the Gadigal had left behind on their forced removal: a kangaroo, a fish, and by a more recent hand, a menacing stick figure in tailcoat and top hat. Contemplating these, even Henry Moseley, his mind tainted by Victorian prejudices, experienced a spasm of regret at the European takeover of this coast a century before that had violently displaced a people so "interesting and original." The only indigenous men and women he would meet on his Australian tour were imprisoned miles inland. There they were denied the opportunity to provide for their own subsistence—including all delicious oyster harvests—and forced instead to tend the cattle and crops of white men.[9]

Eighteen months into their global odyssey, *Challenger* could not be called a happy ship. While Suhm, Moseley, and their fellow scientists enjoyed the privilege and personal funds to explore onshore for weeks at a time, the crew remained bound to an unvarying shipboard routine. When brief leave was arranged for them, they found the class-conscious citizens of Sydney rude and aloof. Stuck onboard through the humid, subtropical winter, it sometimes rained twenty-four hours at a stretch. Below the water line, the atmosphere grew fetid. No one could endure the stuffiness more than a few hours before heading on deck for a proper lungful of air. The rainbow-filled skies and glorious harbor sunsets were considered inadequate compensation.

9 Moseley (1892) 239. The official *Challenger* narrative excerpts Moseley's account of his visit to Berowra Creek in full, with the conspicuous exception of this closing anticolonial sentiment. Australia, the *Challenger* text implies, was terra nullius—a biotic wonderland uncomplicated by people or history.

Worse still, the frustrated officers bullied the hands to distraction. Already in Melbourne, two men had stolen a ship's boat to make their escape; they were now clapped in irons awaiting court martial. On arrival in Sydney, the ship went into dry dock for repairs, then took in reserves of coal and provisions to last them until Hong Kong. This kept the hands busy for a week, but time soon hung heavy on their hands as discontent simmered.

The sailors had a case. Work aboard *Challenger* was harder than on an ordinary man-of-war, with no commensurate increase in pay. They had crossed the Atlantic four times already, and been nearly sunk by an iceberg in the polar sea. And still the widest ocean of all, the Pacific, lay ahead of them. When two more experienced hands deserted, followed by three members of the ship's volunteer brass band, genial Captain Nares finally lost his temper. He blamed his officers and warned them to behave better toward the men. But with centuries of distrust between the two groups, the sailors merely laughed off any gesture of conciliation.

By the time *Challenger* departed Sydney for New Zealand, she was two dozen men short of her complement. For ship's steward Joseph Matkin, who kept a diary, what irked most was the naturalists' casual disregard for the shipboard staff who made their high-minded speculations possible. He read with irritation Professor Thomson's dispatches to the London journals. These trumpeted *Challenger*'s "discoveries" but expressed no appreciation whatever for the hardships they had faced or gratitude to the faceless rotation of naval professionals attending his every need.

In true Victorian spirit, the official narrative of *Challenger*'s Sydney sojourn ignores all personal discontent onboard to focus exclusively on scientific findings, principal among them a magnificent haul of bivalves. A bridgeway between the northern tropics and southern temperate zone, Sydney Harbour possesses the full range of mollusk-friendly habitats: sandy beaches, sheltered rocky shores, mangroves, marshes, mud flats, and quiet estuaries in the style of Berowra Creek. All across the harbor's nooks and byways, the ocean tides mix with freshwater from a thousand rivers, streams, and subterranean springs bubbling up through the porous soil—a true bivalve heaven. For mussels requiring generous allotments of sun, a two-meter intertidal zone courtesy of the Pacific Ocean offers pillowy, moist accommodation.

This wondrous ecosystem—a product of the last glacial retreat—rendered the business of collecting bivalves almost indecently easy. More than fifty species lined the shelves of *Challenger*'s workroom, some of them cosmopolitan types but most wholly new and indigenous to the Australian coast. Several were even dredged from the open sea beyond the heads, though an unbelieving taxonomist in London would later lump these in with the Atlantic mollusks.

What *Challenger*'s success with *Trigonia* fishing and their rich bivalve collection obscured, however, was the imminent collapse of the oyster industry in Sydney Harbour. Captain Cook had found Botany Bay flush with oysters in 1770, and shellfish had been a mainstay of the colonial diet ever since. Oyster bars were popular, accordingly, but the principal economic interest in bivalves lay in construction: shells by the millions were burned in kilns for lime to produce cement, plaster, and whitewash, all necessary raw materials for a booming frontier town.

Alongside *Trigonia lamarckii*, the most historically notable mollusk of *Challenger*'s collection was the so-called hairy mussel, *Trichomya hirsuta*. *Trichomya* dominated the shell middens Henry Moseley stumbled across on Berowra Creek, which were a mere fraction of their original size. *Trichomya* mussels that the Gadigal people had dined on and accumulated across the ages, local lime kilns had decimated in little more than a decade.

Looting shell middens had been insufficient to satisfy demand, however. Live oyster dredging—a form of coastal strip-mining—commenced in Sydney Harbour in the 1860s, extracting hundreds of tons of shellfish annually from the harbor and surrounding estuaries. The crude dredge most commonly used—a heavy iron bar with mesh bag attached—ripped up entire reefs and with them the seafloor substrate that the mollusks relied on to reproduce. Onshore, meanwhile, the clearance of forests for roads, houses, and farms dried up nearby freshwater ponds and delivered tons of silted runoff into the harbor, choking the water column.

In a critical development, the increasingly turbid waters of the estuarine harbor attracted an invasive mudworm from New Zealand: *Polydora*. The tiny parasite took up residence with the oyster in a tube of mud, effectively suffocating it. A diseased oyster was instantly recognized by the telltale blisters covering the inside of its shell. Stillborn shellfish perished in their millions, including entire populations of *Trigonia* and *Trichomya*. In 1886, mudworm disease struck the Hawkesbury River mussel population. A generation prior, oysters had been harvested at the rate of a hundred bags a day from the river. A decade after Henry Moseley's daytrip to Berowra Creek, no shellfish remained.

Already by the time of *Challenger*'s visit, the alarm had sounded over the rampant destruction of Sydney's oyster reefs. A parliamentary Act of 1868—the first regulation ever applied to a Southern Hemisphere fishery—vowed to curb "the criminality of this shameful waste." To no avail. Two years after *Challenger*'s departure, a Royal Commission was still lamenting the ruinous decline of molluscan beds up and down the coast, including the decline of the once plentiful *Trigonia*.

New restrictions were put in place, but nothing could save the oyster beds of coastal New South Wales, which have never recovered to their precolonial state. This is true of temperate, oyster-friendly estuaries across the world, from the coasts of Europe to the Atlantic seaboard of the United States. Mechanical dredging was first introduced to Chesapeake Bay in the 1870s. Today, only 2 percent of the bay's native population survives, while more than 85 percent of shellfish reefs worldwide are "functionally extinct." Aquaculture, not nature, produces oysters for the modern world's dinner tables.[10]

Popular ignorance of this fact is an example of shifting baseline syndrome, a "generational amnesia" whereby the existence of entire healthy ecosystems, once destroyed, can be universally forgotten in a matter of decades. As early as the public debates of the 1890s in colonial Sydney, "the question arose as to what a natural oyster bed *was*." Each successive generation of human beings presumes the world they enter to be what it has always been. Dramatic systems-level change—such as the disappearance of the world's natural oyster beds—is too fanciful to imagine.[11]

After a two months' stay in Sydney—and laden with pretty *Trigonia* souvenirs—HMS *Challenger* left the sanctuary of the famous harbor to the cheers of the assembled locals. They cruised toward the heads, careful to avoid the breakers crashing over the Sow and Pigs reef. But the Tasman Sea greeted them with a fierce gale, forcing *Challenger* to beat a retreat to Watsons Bay.

When the winter storm abated, they resumed their appointed task of sounding the ocean floor between the Australian east coast and New Zealand to prepare the laying of a telegraph cable. It would be no easy job. Their soundings confirmed the

10 Ford and Hamer (2016) 87.

11 Alleway and Connell (2015) 795; Thomson (1893) 62 [emphasis added]. "Much of the ecology of eastern Australian estuaries over the past millennia was arguably dependent on the reef-forming habit of oysters. The ecological impact of the permanent and complete demise of subtidal oyster reefs in these estuaries is unknown." Ogburn (2007) 282.

abyssal depths of the thousand-mile-wide Tasman Sea. New Zealand was no breakaway peninsula of the Australian continent but an independent landmass separated by churning, powerful currents. With storm after storm and heaving seas, it was one of the roughest stretches of ocean they encountered the entire voyage. Ten miles out of Wellington Harbour, a freak wave caused havoc in the mess and washed a man overboard. By the time anyone noticed, it was too late.

During breaks in the weather, the Challengers resumed their practices of regular soundings and temperature readings. With this original data for the Tasman Sea at hand, the larger western Pacific Ocean system, of which it was a part, came into focus. Thirty miles southeast of Port Jackson, they met with a powerful coastal boundary current channeling volumes of tropical water southward. The surface water temperature, which had been 62°F (16.7°C) in the harbor, rose to 70°F in the heart of the current. The entire southwestern Pacific, the Challengers ultimately determined, was characterized by this "striking extension southward of [warm ocean] temperature."[12]

The so-called East Australian Current (EAC)—first identified by James Cook a century before *Challenger*—snakes from Sandy Island off Queensland southward along the New South Wales coast. It then splits in two—one branch tailing eastward to New Zealand, the other bringing residual warmth to Bass Strait north of Tasmania. When, after a thirty-mile traversal, the sea surface temperature abruptly dropped, the drag of the current on *Challenger*'s hull vanished with it.

The high temperatures of the EAC directly diminished *Challenger*'s haul with the tow net and dredge. Offshore warm waters from the north, the philosophers guessed, were nutrient poor, depressing the volume and diversity of marine life in contrast to the cooler, less turbid waters of Sydney Harbour with their mollusks in dizzying abundance.

12 Buchan (1895) 15.

Today, the EAC is a poster child for climate change impacts on ocean circulation, and its high temperatures are a threat to marine ecosystems along the Australian seaboard. Pumped with extra heat, the "intense, narrow, poleward boundary current" is at least 20 percent more powerful than a half century ago. Moreover, the EAC has extended its reach southward some 220 miles (350 kilometers) beyond its historical limit at the northeast tip of Tasmania, now bathing the southern coast of that island with warm water drawn from the faraway tropics.[13]

This enhanced EAC belongs, in turn, to a larger oceanographic phase shift. The so-called spin-up of the South Pacific gyre is now impacting thermohaline circulation across the Australasian region, generating enormous subsurface bands of heat. Seasonal extreme events, so-called marine heat waves, pose a threat to entire undersea communities of southeast Australia. The island state named for the Dutch explorer Abel Tasman is most vulnerable of all. While westerly winds maintain the prevailing current at a distance offshore of New South Wales and Victoria—insulating coastal marine life from heat and invasive biota—the east coast of Tasmania lies in the direct path of a steroidal EAC. Waters there are heating up *four times* faster than the global average—the hottest of hotspots in the Southern Hemisphere—with devastating consequences for its native kelp forests.[14]

It's possible the decimation of coastal oyster beds in the days of HMS *Challenger* undermined the long-term resilience of the keystone kelp forests of Tasmania. In her winter crossing of the Tasman Sea in 1874, *Challenger*'s bow sliced through thick ribbons of seaweed on the ocean surface. A century and a half later, that kelp has mostly disappeared. Increasing ocean temperatures have laid waste to its native inshore habitats.

13 Ridgeway and Dunn (2007) 4.

14 Roemmich (2007) 162; Oliver et al. (2014) 1980.

Overfishing of keystone herbivors has deprived the kelp forests of their traditional caretakers, inviting disease. At Bruny Island, where François Péron collected the first living *Trigonia* shell, the precolonial undersea kelp stands are now 98 percent vanished. This has precipitated "vast changes" to the marine ecosystem of the Tasmanian east coast. Hundreds of algae and invertebrate species, some first identified by *Challenger*, have disappeared, along with the keystone seaweed and three dozen native fish species.[15]

Into this vacuum created by the kelp crisis, an invader from the north has seized the opportunity for territorial expansion. Migrating south at the rate of 100 miles (160 kilometers) a decade, *Centrostephanus rodgersii*—a rapacious sea urchin with the "ability to catastrophically overgraze seaweed beds"—has transformed the ancient, kelp-rich coastal reefs of the Tasmanian north coast into "barrens," an unrecognizable ecosystem characterized by impoverished reefs, invasive tropical species, and drastically lowered diversity. Incipient colonies of the destructive *C. rodgersii* have now been spotted off Tasmania's south coast. The immediate biological future of this once richly diverse marine zone is, in a sense, utterly predictable. Faced with the twin threats of warming and new biotic aggressors, and with no attainable land mass to the south, the marine wildlife of the Tasman Sea faces mass extinction.[16]

This potentially includes a multitude of mollusk species. Southeast Australia boasts one of the richest mollusk populations in the world, *Trigonia*, *Trichomya*, and a thousand others having radiated outward from tropical seas over tens of millions of years. Geographical isolation since the final breakup of the supercontinent Pangea and the creation of the polar oceans means that 95 percent of Australian mollusk species are endemic.

15 Ling (2008) 891.
16 Ling (2009) 720.

But in just fifty years—a blink in Earth's history—warm currents have returned to the temperate mollusk habitats of the Tasman Sea, threatening mass mortality. Warmer water depletes oxygen and raises mollusk metabolic rates to fatal levels, inhibiting growth and narrowing the window for successful propagation. More dangerous still is the increasing acidity of the ocean water, which reduces the availability of carbonate minerals for shell formation.

The death blow, conceivably, will be the demise of the caretaker kelp forests. A recent study has shown that nearshore kelp stands provide a daytime refuge from acidic waters and potential deliverance for shell-building invertebrates in acidifying oceans. Tragically, today's historically elevated sea temperatures are just the beginning. A further 3°C (5.4°F) of warming predicted for 2070 will push almost all native marine fauna outside their climate envelope. And with Australian coastal reef systems stripped of kelp, those local ocean creatures incapable of locomotion, clams and mussels among them, will have nowhere to hide.

Four months after their Sydney visit, the *Challenger* arrived in Cape York in northern Australia, where the naturalists observed the indigenous residents harvesting mussels from the beachside shallows. A century later, anthropologist Betty Meehan spent a year with the Gidjingali people on the nearby shores of the Arafura Sea—once a land bridge connecting New Guinea to Australia—where the naturally self-replenishing oyster population continued to sustain more than a dozen families. Many of the bivalve species that *Challenger* collected were still flourishing in Arnhem Land according to her 1970s account, including *Trichomya* and the charismatic *Trigonia*. In coastal regions free of a colonial footprint, Australia's oyster reefs have endured—until climate change.

Indigenous Australians have collected pearly shells for millennia to fashion for personal adornment. In that sense, the craze for "brooch clams" in colonial Sydney merely mimicked ancient local practices. Decorative shell traditions have persisted, generation to generation, among the First Nation communities of Australia in the centuries since European invasion. In a different sphere of cultural memory entirely, *Trigonia* was for nineteenth-century Europeans a proof of Darwinian continuity in the history of life. With climate change, however, the modern variants of that legendary fossil now reemerge as figures of loss, extinction, and the Anthropocene. Along a superheated Australian east coast, the brooch clams, hairy mussels, and their kin are disappearing from view, to be interred with their forebears in the vast midden heap of extirpated life, terrestrial and marine—once and future fossils in the inglorious landfill of the sixth extinction.

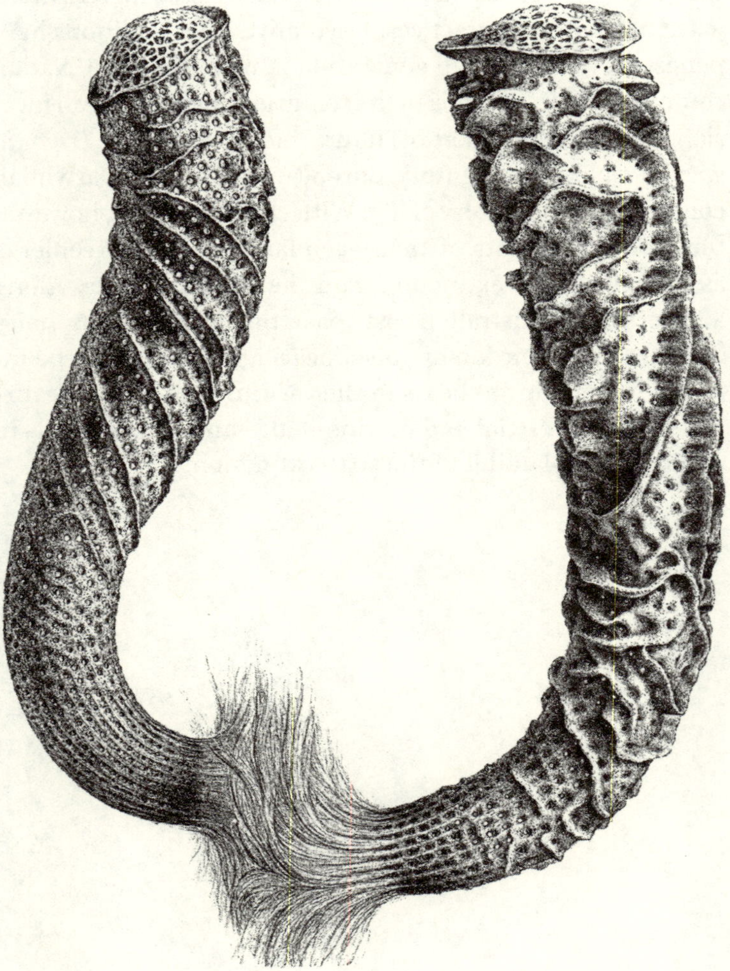

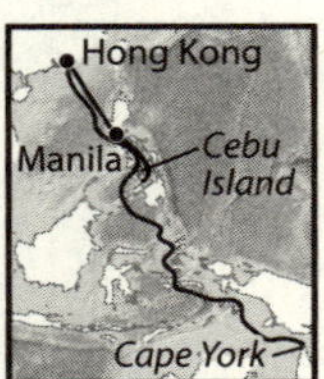

FIGURE 8.1 The fabled Venus Flower Basket sponge (*Euplectella aspergillum*). From *Scientific Results: Zoology*, vol. 21. plate 1 (1887). (Inset) HMS *Challenger* Leg 8. Cape York to Hong Kong to the Philippines. September 1874–January 1875.

CHAPTER 8

The Recklessness of Beauty

Cebu Island, The Philippines
10°19′ N, 123°45′ E

Wyville Thomson was a portly, bearded gentleman who would not have looked out of place onstage in a Gilbert and Sullivan operetta. But like many a repressed Victorian, he felt a deep affinity for beautiful objects. His tastes were unconventional, however. Marble statues celebrating the human form did not satisfy his aesthetic craving; neither did enameled snuff boxes. Instead, he was drawn to examples of animal elegance the most remote—in space, time, and form—from his own person. Nothing answered this criterion better than the deep-sea sponges—those sheer, bodiless relics of Cambrian times looted from the ocean's abyss, whose response to the exigencies of existence could not be more different from the lubberly arrangement of arms and legs he surveyed each morning in the mirror.

The glass sponges spoke particularly to Thomson's platonic ideal. Called Hexactinellida, and first identified by Thomson, these unique sponges are possibly the most ancient creatures still living—some 800 million years old. "They are," accordingly, "of unique interest from almost every evolutionary point of view." The glass sponge *Hyalonema* sports threads of clear silica woven together in the form of a coiling rope, crowned with glistening spun glass. Karl von Siebold—mentor to Willemoes-Suhm—had first brought a specimen back from Asia to universal amazement. "Nothing had been seen in the least like it before," Thomson remembered. But a different genus,

Euplectella of the Philippines, represented the acme of glass sponge architecture in Victorian eyes. One species, *Euplectella aspergillum*, had earned the nickname Venus Flower Basket because its woven tubular form lures mating shrimp inside. The Japanese name for this iconic sponge—*kairou-douketsu*—strikes a similar theme: it means "together for eternity." For a shrimp couple imprisoned for life beneath the flower basket's lattice lid, monogamy is guaranteed.[1]

On a foggy September morning in 1868, off the Faroe Islands north of Scotland, Thomson had lucked upon a crowded bank of never-before-seen glass sponges. One emerged as a perfect, gelatinous sphere with threads in a star pattern across its surface and a maze of glassy filaments for a base. Though delicate-seeming, the sponge meshwork was unusually supple and strong. He named it *Holtenia carpenteri* after his dredging companion on HMS *Lightning*.[2]

Five years later in deep waters west of Gibraltar, Thomson, now aboard *Challenger*, came a tantalizing step nearer the holy grail. Several tubular sponges, complete with the familiar glass latticework, tumbled on deck—albeit torn by the clumsy dredge. The first *Euplectella* to arrive in Europe had been mistaken for an exquisite Chinese vase crafted by human hands. Here was evidence of a new species of flower basket sponge thriving a world away from its Philippine nursery.

The deep-sea Hexactinellida—that "beautiful family"—spanned both time and space. Glassy imprints studded the chalk fossil beds of south England. But could the "rich, imbedded lettering" of the limestone be interpreted correctly without an intact modern counterpart for reference? By the time of *Challenger*'s departure from England in December 1872, Thomson's desire for a living flower basket sponge had become almost an obsession—a circumstance well-known to

1 Leys et al. (2007):132; Thomson (1867) 31; Saito et al. (2002) 521.
2 Thomson (1873a) 73.

the ship's officers and naturalists alike. The deep waters off Cebu Island, home of the *Euplectella*, had been circled on every map from the first day.[3]

Time, the ultimate craftsman, labors best in obscurity. For the Victorians, as for us, there is no place more remote from human observation than the ocean's abyss. The assumption was, and is, that in the deep sea lie answers to the great questions of our planet's formation and life on Earth. For Wyville Thomson, therefore, sailing *across* the ocean represented a horizontal means to a vertical end. *The Depths of the Sea* was the title he chose for his famous book; it was the oceangoing philosophy he lived by. The average seaside, gaudy with sunlit corals and shells in easy reach, seemed to him mundane compared with the oceanic depths and its peculiar aliens. Only there was that "recklessness of beauty" truly apparent "which produces such structures to live and die, forever invisible, in the mud and darkness of the abysses of the sea." *Challenger*'s chief scientist was an empirical man with a bent to infinity and its earthly signs: beauty, space, and time. All were united in the sublime prism of the deep-sea sponge *Euplectella aspergillum*.[4]

Challenger's passage to the Moluccas led them through some of the oldest European settlements in Asia and the most biodiverse seas on Earth. At Ambon Island, entrepreneurs of the centuries-old spice trade had cleared the native forests for plantations of clove and nutmeg and enslaved its indigenous peoples. The harbor itself—indifferent to human travesties—offered the most beautiful spectacle imaginable for an oceangoing philosopher. Corals, sponges, and anemones in fanciful shapes and colors lay shimmering in waters translucent as air. The seafloor, fifty feet down, was uneven, studded with rocks,

3 Thomson (1877) 1:141; Thomson (1869) 124. Hexactinellids are in fact global in their deep-sea distribution, as *Challenger* established.

4 Thomson (1869) 120.

chasms, and little hills and valleys where the miniature animal forests grew. Coral fish, banded and spotted blue, yellow, and red, darted among the reefs. Rose-colored medusae vacillated serenely above them.

Further north, the Philippine archipelago hosts three-quarters of the world's corals and almost half of all known fish species. Strolling along the coralline beaches, the Challengers collected buckets of ornate shells, many with established market value among European collectors. In crystal, pure, shallow waters, the reefs presented an orgy of reds, purples, oranges, and cerulean blues visible from the small boat. Coral fragments emerged from the trawl in all shapes: spiky and tubular, some elegant as bouquets, others squat like toadstools.

Heaped among the corals on *Challenger*'s deck was a cabinet's worth of novel starfish, sea cucumbers, spider crabs, and mollusks. Some they recognized, like the "jewel box oyster" recently collected in Fiji, and the twisted "moon snail." Most remarkable of all, arguably, was a phosphorescent tunicate shaped like a human heart, complete with purple arteries pulsing light. At Hong Kong, over Christmas, they had packed no fewer than 129 crates of specimens to mark their first two years at sea—special delivery for London. Already replacement crates in the hold were filling up.

They also shipped a new captain to replace George Nares, who had been reassigned to an Arctic expedition, in addition to sixty new hands ordered from London to fill their ranks depleted by desertions. At the news of Nares's departure, Wyville Thomson flew into a rage and threatened to cancel the entire enterprise. But he soon faced the brute naval reality. There was no one to listen to his complaints—and they were still nearly two years from home. Captain Frank Thomson (no relation) was a hard-drinking, garrulous Navy man, who arrived on board accompanied by his piano, a cello, and a brawling cat.

On their return to the Philippines, south of Masbate Island, they passed a nest of writhing sea snakes striped black and yellow—these were considered too dangerous to catch. Then the wind failed them completely. Resorting to steam power, *Challenger* chugged amid a cluster of small, wooded islands rimmed with dazzling white beaches. They came at last to the narrow strait separating mountainous Cebu from the low-lying coral island of Mactan. Here Ferdinand Magellan had been killed during a beach assault in 1521. His death had not forestalled Spanish annexation of the region, and the *Challenger*'s new captain sent the ship's photography team to record the memorial erected in Magellan's honor. Like the conquistadors of old, the Challengers were compelled to leave their boats a distance from the shore and wade across the wide coral reef, photographic apparatus held high.[5]

By Victorian times, trade at Cebu was mostly in British and American hands, not Spanish. *Challenger* berthed alongside a Liverpool coaler out of Sydney bound for China. Other vessels carried locally grown tobacco and sugar for European markets. On Cebu, a year's worth of cigars could be bought for an old knife.

In the lucrative sea sponge trade, however, the native Visayans held the advantage, at least for a time. Only local fishermen knew the location of the precious *Euplectella* bank. They had also developed a custom-made bamboo dredge to raise the sponges intact. When a European sponge-seeker came to port, he was invariably pointed in the wrong direction. But the local monopoly hadn't lasted. When the waters in the Cebu channel had first been harvested ten years prior, the mysterious glass sponges fetched the modern equivalent of four thousand dollars apiece. Now, with the Venus Flower Basket under glass in museums and drawing rooms across Europe, the going price had crashed to a little over ten dollars.

5 *Challenger*'s photographers, who had a tendency to desert, left a rich visual record of the expedition but almost exclusively of scenes and peoples onshore.

On a sticky morning in January 1875, the late-coming philosophers of HMS *Challenger* fulfilled at last their destined appointment with the glass sponge *E. aspergillum*. Cebu boasted the richest vegetation they had yet seen. Bamboo thickets and palm trees contested every square foot of ground in the town, even crowding the sky. Thomson, accompanied by Moseley and Murray, set out early in the pinnace for a fishing village to the south. The heat was already oppressive, but the mosquitoes, at least, had given them a morning's respite. Trees near the shore were thick with flying foxes, their heads hung down, apparently sleeping. On the beach, flocks of curlews and the graceful white egret stood among the fishing boats. Here Thomson enlisted two local guides for a pittance, with orders to bring them to the *Euplectella*.

Of the dazzling variety of mollusks the Challengers trawled along the shores of Cebu in early 1875, many were already familiar to them from the multivolume *Conchologica Iconica* (1843–1878), which featured the shell collections of Hugh Cuming, first British naturalist in the Philippines. Cuming had enjoyed the hospitality of the Spanish government at Cebu in the 1830s and 1840s. He sent the local schoolchildren out in search of natural wonders, while he complained of an irritated bowel and "eyes . . . much injured by the sun."[6]

At first, the Visayans had refused to believe Cuming's mission was to collect shells, snails, and sponges, or that a faraway market existed for them. He thus probably paid next to nothing for the *E. aspergillum* he sent to the anatomist Richard Owen in 1841, the first seen in England. But Owen understood immediately the value of his prize—"the most singular and beautiful, as well as the rarest of marine productions." Owen described for the Zoological Society of London a truly unique object: a skeletal lattice in the shape of a horn

6 Dance (1980) 487.

made up of shiny longitudinal, transverse, and oblique fibres "resembling the finest hairs of spun glass," winding spirally in opposite directions. On top of this mesmerizing production sat a tufted lid like an Elizabethan ruff. Someone at the Society suggested the shrimp inhabitants of *Euplectella* had built this fairytale residence for themselves. He was roundly mocked, even if no one had a better explanation for how this wonder of nature had come to be.[7]

Wyville Thomson, who first saw Cuming's *Euplectella* at the British Museum, contemplated the new order of glass sponges through the lens of deep time. The "Ventriculites" of the English coast—tubular, vase-like fossils with grooved decorations on their surface—had long confounded paleontologists. Now, here was a modern descendant to spotlight Thomson's pet theory of the continuity of the chalk. "We are still living in the Cretaceous period," he announced, again to general uproar. Thomson's formal analysis of *Euplectella* would prove more enduring; it anticipates glass sponge research of the twenty-first century carried out with electron microscopes and 3D printers. He approached the task with caution because previous accounts had been riddled with errors: "There seems to be something unusually fatal," Thomson reflected, "in the fascination of these beauties." The enigmatic perfection of the Venus Flower Basket defied all attempts at description.[8]

First, he was surprised at the sponge's toughness, which belied its seeming fragility. *Euplectella* could withstand rigorous inspection and handling. His microscope—limited as it was—disclosed keys to this resilience. Each glass fiber (called a spicule) contained a thin axial tube, ensuring suppleness and integrity. These individual glass threads were then cemented together in an elaborate, curved latticework—like the interior ceiling of a classical dome—the object of which must be to

7 Owen (1841) 3–4.
8 Thomson (1871a) 225; Thomson (1868) 127.

absorb the ceaseless buffeting of currents when fixed to the ocean floor, while also allowing nutrient-rich water to filter through it. The whole was an engineering marvel, a masterpiece of adaptation to the deep-sea environment. The recklessness of beauty? More the resilience of beauty.

For all the hundreds of Philippine glass sponges now in private hands in Europe—and the voluminous commentary upon them—a living Venus Flower Basket had yet to be properly described. It was with a desire for this singular trophy that Thomson and his colleagues approached the *Euplectella* ground off the coast of Cebu. In the eyes of their Visayan guides, the sponge resembled the spout of a watering can; they called it *la regadera*. They had brought with them onboard their ingenious trap made of bamboo, assembled in the shape of a *Y* and armed with fishhooks. They directed the Europeans to a nondescript area of open water three miles off shore. Conditions were ideal: a full tide, smooth water, and a strong undercurrent to maneuver them steadily across the quarter-mile-wide bank.

The seafloor here was made of mud and sand, said the guides, with the base of *las regaderas* anchored in the soft sediment, and its oval mouth directed *a donde se pone el sol* (toward the setting sun). Thomson and the others watched the Visayans lower the dredge into the water on a thin hemp line to about a hundred fathoms. After an hour's combing the seabed, they pulled the trawl gently to the surface. The moment it emerged was charged with excitement for *Challenger*'s naturalists, who counted three intact glass sponges attached to the dredge. With its perfect architecture honed by half-a-billion years' evolution, an individual glass sponge may live for millennia—until it is rudely extracted by a fishhook on a bamboo rod.

E. aspergillum begins its life as a free-swimming larva, before attaching itself to the rocky bottom and metamorphosing into an adult sponge. On board, still alive, Thomson's *Euplectella* remained coated in a brown, gelatinous slime—its living protoplasm. From what can be deduced, the Venus Flower Basket subsists on organic matter sunk from the water's surface, which it transports through a network of tissues as if it were a plant. These same tissues carry electrical signals analogous to an animal nervous system. The glass sponge's antiquity and uniqueness are in close correlation.

Under Thomson's probing fingers, the silicate harness beneath the sponge protoplasm felt rigid but flexible. On return to shore, the Visayan fishermen hung their catch on a tree branch in the sun to bleach. After an hour, the protoplasmic casing dried and fell off, revealing the ornate skeleton the philosophers had journeyed thousands of miles to see. Thomson then examined for himself, at last, in the wild, the unique silicate design "whose beauty cannot be added to nor taken from." The sponge was more brittle in its naked state but still surprisingly tough.[9]

By day's end, they had brought washing baskets full of *Euplectella* back to the ship, purchased for two shillings a dozen. In his diary that evening, Suhm reflected on the expedition's banner day at Cebu: "That Prof. Thomson himself handled the sponges with the utmost care, freshly examining them in all possible fluids as well, I need hardly mention." Suhm counted four inmates of the luxury glass prison: a see-through shrimp, a white worm, an isopod, and a tiny clam. Despite their success, the professor's ardor for the glass sponge was not yet satisfied. But when, three days later, *Challenger* returned to the very same spot in the Cebu channel to trawl for more *Euplectella*, they got nothing. Within two years of their visit, the seabed off Cebu was completely barren of sponges.[10]

Sponges—simplest yet most enigmatic of marine animals—had inspired in Thomson's fellow Scotsman Robert Grant a visionary theory of evolution for which he paid a ruinous

9 Chimmo (1878) 11.
10 Willemoes-Suhm (1873–1876) Letter LXXV.

professional price. If the sun-warmed Aegean Sea marks the classical birthplace of European marine science, its Victorian-era renaissance occurred at the continent's western margin in the chilly waters of Scotland's Firth of Forth.

In the 1820s, the University of Edinburgh, a brisk walk from the estuary's beaches, was home to an influential postwar generation of free thinkers, Grant prominent among them. Unwelcome at conservative Oxford and Cambridge, these Scottish radicals advanced marine biology as a vehicle for new, secular ideas about the history of life and, by extension, human origins and purpose. Grant mentored Charles Darwin in his Edinburgh heyday but died destitute and forgotten the year HMS *Challenger*, whose trail he blazed, crossed the Pacific Ocean.

With the teenage Darwin occasionally in tow, Grant spent the winter of 1826 gathering sea pens, slugs, and sponges from the shallows of the Firth of Forth. Braving sleet and numb extremities, he lay prone on slimy rocks at low tide to observe the mysterious rituals of sponge behavior. He crawled into caves to extract the "downy fleece" hanging by the thousands from the roof. At his improvised laboratory in the bay at Prestonpans, Grant kept dozens of sponges alive in tubs. He was mesmerized by the "impetuous torrent" of liquid waste sprayed from their holes. He wondered at their irregularity of size—from half an inch to four feet broad—at their rainbow range of colors, and their bewildering variety of odors (some smelled like mussels, some like mushrooms, others like nothing at all). He tracked their free-swimming larvae and pondered their density of population—sufficient to gum up the dredges of Firth fishermen—as well as their remarkable geographical spread from the icy shores of Greenland to sunny Greece to distant Tahiti. At the University of Edinburgh museum, he exhibited the wonders of the sponge before a rapt audience of would-be Lamarckians. He argued that only a progressive theory of life—species shading one into another over vast stretches of time—could explain such stunning ubiquity and variety in a single creature.

In a landmark suite of papers on the "philosophy of the sponge," Grant coined its phylum name (Porifera) and speculated on "the great purpose which it is destined to fulfil in the universe." Sponges were some of the earliest inhabitants of the world's oceans, as attested by the fossil record. A simple freshwater sponge (still common across Europe) was presumably ancestor to its marine progeny, which had acquired greater complexity "during the many changes that have taken place in the composition of the ocean." The primitive sponge's canals, he surmised, were the earliest iteration of a stomach. Everywhere the sponge went, marine life flourished, proof of its keystone role. The Porifera offered a labyrinthine shelter for innumerable minute creatures and precipitated harmful carbonate from seawater, "purifying the vast abyss" of the global ocean for more advanced life.[11]

Grant's French-inspired philosophy of sponges made a powerful impression on his Edinburgh circle, including Edward Forbes. From Grant, the *Challenger* philosophers inherited a core evolutionary principle—originated by Lamarck—that species transmuted over time in progress toward ever greater complexity. From Lamarck's compatriot Geoffroy Saint-Hilaire, moreover, with whom Grant studied in Paris, they reckoned these changes were triggered by directional changes in the marine environment, most notably global cooling. Climate change had driven sponge evolution, for example, just as over geological time it had turned some cephalopods into fishes.

Global measurements of ocean temperature, water chemistry, and depth are hallmarks of the fifty-volume *Challenger* reports. The temperature data alone exceed 5,000 individual readings. These efforts reflected the Edinburgh school's

11 Grant (1825–1826) 95, 99, 136; (1826) 283.

preoccupation with environmental conditions, its Lamarckian legacy. In the spirit of their forerunner Robert Grant, the *Challenger* scientists spent the long days at sea huddled over microscopes, dissecting and drawing, and nights disputing the subtleties of evolution. The marriage of environmental data with morphology—study of an animal's physical development through its life cycle—would reveal the place of *Challenger*'s myriad specimens in the history of creation, or what the atheist Grant insisted on calling the "economy of Nature."[12]

Challenger's flower basket sponges from Cebu ended up in the hands of German anatomist Franz Schulze, who published his definitive account of the genus in 1887. Schulze's breakthrough insight was that *Euplectella*'s signature checkerboard pattern consists of two interwoven but independent lattices, the horizontal wrapped within the vertical grid. This combination of fusion and freedom is crucial to the overall resilience of the sponge because it allows the stress of ocean currents to be diffused through the structure. Unlike a factory-made car windshield, a crack in one place on the glass sponge need not be catastrophic to the whole.

Nature excels in resilient structures built from inherently brittle materials such as glass. The *Euplectella*, object of fascination to the Victorians, retains its aura today. Twenty-first-century engineers in quest of the next generation of cheap, lightweight, and durable product designs have focused on the latticework of the flower basket sponge with its 500-million-year record of high performance. Glass "spicules" it turns out—*Euplectella*'s basic building block—closely resemble human-made commercial telecommunication fibers, only longer-lasting.

A suite of recent research papers has found that the Philippine glass sponge embodies resilient engineering at every

12 Grant (1826–1827) 137–38.

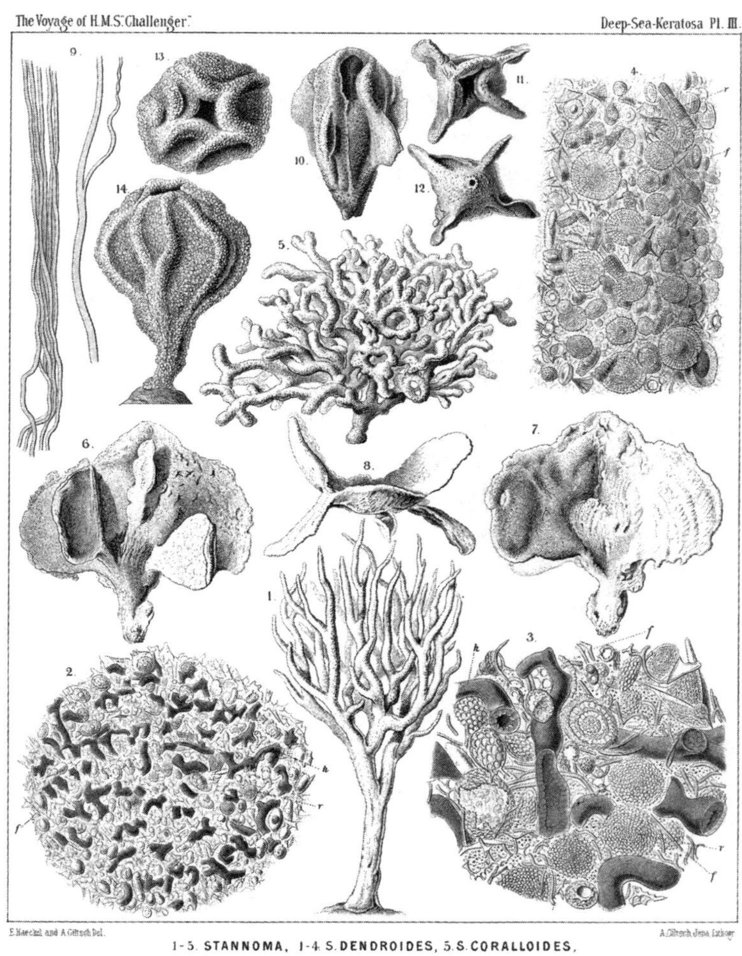

FIGURE 8.2 Ernst Haeckel's rendering of deep-sea sponges collected by *Challenger*. From *Scientific Results: Zoology*, vol. 32, plate 3 (1889).

level, from the laminated silica nanoparticles of its individual threads—which allow the stressed spicules to crack without breaking—to the counterintuitive pattern of woven diagonal struts that skip every second square cell. Diagonal reinforcement of *all* cells would require more material for no advantage in strength. Because sponges, like humans, live in a world of strictly limited resources, evolution cherishes efficiency above all things. A 2019 experiment with 3D printing a Venus Flower Basket—allowed only the same material amounts as in nature—failed to replicate the sponge's strength, or its beauty. Industrial biomimicry of the Victorians' beloved *Euplectella* remains "aspirational."[13]

The glass sponge fascinated Wyville Thomson with its inimitable elegance but also its antiquity. *Challenger*'s work stands as the founding taxonomical research in the field. In a public lecture in Glasgow he gave soon after *Challenger*'s return, Thomson speculated on the existence of charmed oceanic zones—the Faroe Islands, the coast of Portugal, the Philippines—which had by chance been preserved through the geological upheavals that had transformed the rest of the planet and existed as de facto museum cabinets for "strange and beautiful things" remnant from Earth's past. The *Euplectella*, for example, through some combination of luck and resilience, offered "a glimpse of the edge of some unfamiliar world." Thomson's writings on the life of the ocean abyss are full of romanticisms of this kind.[14]

More than luck, however, has sustained the Porifera through five mass extinctions in Earth's history—and primed it to thrive in the sixth. As for the glass sponges, despite a worldwide presence, and some 500 distinct species, much about them remains mysterious, including their reproduction. The greatest impediment to study—as the Challengers discovered that hot morning off Cebu Island in 1874—was in keeping their specimens alive. Once disturbed from the seabed, basket sponges are all too ready to shed their living tissue and assume the otherworldly postmortem appearance for which they are famous.

13 Robson Brown et al. (2019) 11.
14 Thomson (1876a) 495.

Thomson admired the simplicity of the glass sponge body plan and its ability to nourish itself through absorption of organic matter in the water in which it is immersed (the sponge is not a hunter after prey—that was a later evolutionary twist). That said, can the glass sponge's ancient mechanism for secreting silica rather than carbon from the ambient water column be counted as pure luck? Whatever the case, this feature protects the Hexactinellida from the acidifying waters that spell doom for their principal competitor on the world's reefs, the carbonate-bodied corals. With habitats cleared of corals, a near-future, less biodiverse ocean might well be crowded with sponges, though of significantly fewer varieties. Will the *Euplectella aspergillum*, first famed for its loveliness under glass, be among them, or will it be consigned permanently, like any other Victorian collectible, to the desiccated afterlife of a museum exhibit?

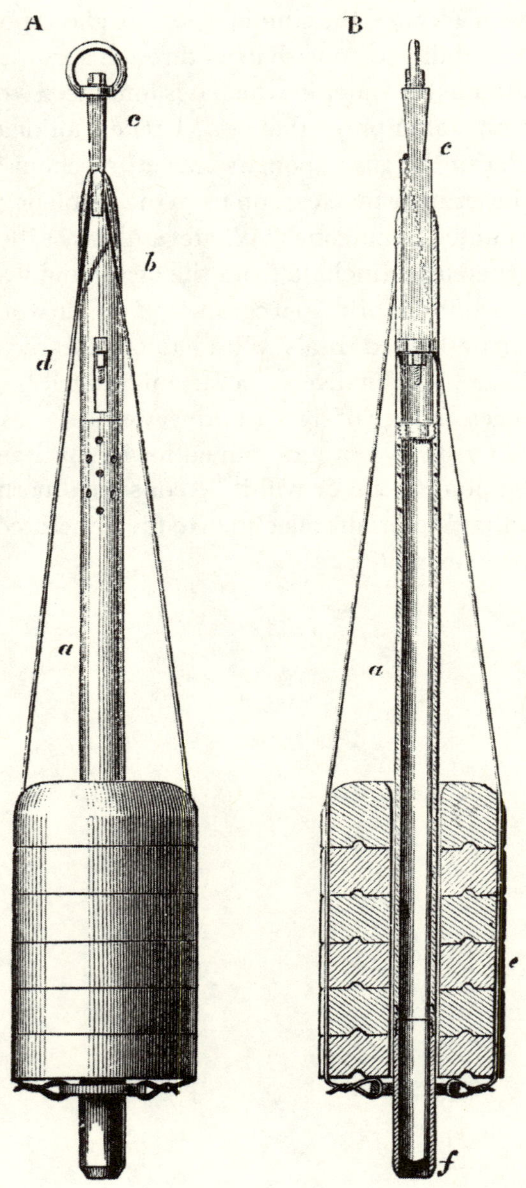

FIGURE 9.1 Sounding apparatus design for HMS *Challenger*. From *Scientific Results: Zoology*, vol. 32, fig. 14 (1889).

CHAPTER 9

The Challenger Deep

Southwest of Guam
11°22′ N, 142°35′ E

Sailing south from Cebu, the Challengers came to an island volcano bearing marks of a recent eruption. One flank of the mountain forest was sunlit green, the other withered by sulfurous gases from the still-smoking crater. Only 1,000 of the 25,000 pre-eruption inhabitants of the island remained, they were told. Later, Moseley and Suhm set out in a local's well-armed canoe to search for the giant coconut-eating crab, *Birgus lato,* along the pink, coral-rich shores of Grande Santa Cruz island. They encountered neither the much-feared Sulu pirates nor the crab, whose beach habitat had been ravaged by invasive pigs.

But in deep waters off nearby Miangas island, *Challenger*'s trawl delivered a fantastical bounty: more than twenty unique fish and a hundred slimy invertebrates new to science. A parade of brittle stars, sea urchins, and a novel barnacle emerged, appearing under the microscope folded like a bishop's mitre. More stunning still were new, strange denizens of the deep. They had drawn blind lobsters and spider crabs from the abyss before, but a cutthroat eel, a wicked-looking hammerfish, another with a rat's tail, and a black lantern fish, with its lightbulb lure set out in front of its hideous jaws, raised the image of a deep-sea world entirely separate from their known world—a dark dreamscape of evolution.

Nature had balanced its ugly-looking fish with a bumper crop of elegant sea lilies: five new genera and a dozen new species, duly named for Thomson, Murray, and Moseley. Ernst Haeckel later rendered two of these deep-sea crinoids as extravagant tassels on a stem, like an expensive feather duster. Then, the weirdest catch of all: a bulb-shaped octopus with elephant ears for flippers, giant eyes, and arms flailing like fronds from a fern pot: *Grimpoteuthis*, the so-called dumbo octopus.

Challenger had begun its journey through Southeast Asia according to guidebook recommendations—in the comfortable dry season. They watched the sun rise above the island palm groves and complete its fiery descent into the ocean out of a blue sky unfiltered by cloud. Warm, dry winds blew from the east. But as their journey meandered into February, the summer heat intensified. Then the intermittent breeze faded to nothing. One morning, a tame parakeet that had lived for months in the rigging dropped dead on deck. Torrential rain flooded the scuppers day after day until chemist John Buchanan reported a measurable freshening of the surrounding seawater.

Captain Thomson abandoned his plans for more northerly explorations. Instead, he allowed the ship to drift south and east on the current. Deprived of the motive energy of the trade winds and low on coal, they now entered waters empty of ships, even of pirates. Beyond the Spice Island trade zone, they were among the first Europeans to encounter the peoples of Humboldt Bay, New Guinea, who surrounded *Challenger* with armed canoes. The New Guineans regarded the invaders warily and would not permit the naturalists to collect specimens on shore. During a botched attempt at bartering for souvenirs, one aimed a drawn bow directly at Suhm, who felt a distinct tingle down his spine. The Challengers had brought guns but thought better of using them.

Here, where the Europeans and Chinese had not yet introduced their plows and enslaved laborers, the island forests of the western Pacific maintained themselves through growth and decay in unthinking splendor. Once the purple line of the New Guinea coast faded from view, *Challenger* found herself in a seaborne forest. Lines of driftwood, some entire trees complete with roots, clung about the ship's sides. Lowering their nets, Murray and Moseley gathered fruits and seeds floating between the bits of trunk, along with mollusks, a rich scum of radiolaria and diatoms, a slate-colored crab, and a writhing, black-yellow sea snake. No sign of leaves or other foliage, however, which had already sunk to the bottom nearer the unseen shore.

The map assured them New Guinea still lay beyond the horizon to the southeast, at that point on its coast where the great Sepik River runs into the sea. An awesome terraqueous cycle was evident. The river had brought both shoreline mangrove trunks and the decayed trees of the inland mountains miles to the coast, then set them adrift on the sea. Much of the wood was hardly floating at all but hung vertically suspended. Gannets and boobies swooped and settled around the flotsam, while sharks and dolphins in numbers hunted among schools of small fish drawn to the woody nutrient pools, causing a constant splash. Tiny worms swarmed across everything, living and dead. The sea surrounding this chaotic hub of existence took on a glassy, oily appearance.

From the ship's boat, Moseley studied the uneven progress of waterlogging in the weird trunks. One end of each log was, by whatever accident, more susceptible to infusion and so sank first, leaving the other end bobbing at the surface. He was surrounded by hundreds of these doomed poles splashing and scraping against each other like shipwreck victims. Sinking was a process, a progress. He was observing, in infinite slow motion, the absorption of terrestrial life into the ocean, to be returned perhaps, one distant day, as a single fossilized twig in a cliff face.

They didn't know it, but they were nearing the Challenger Deep. First, however, they had to contend with constant, steaming rain and days of baffling winds. During their weeks-long northward passage to the Sea of Japan, the wind never exceeded six knots. More often they were becalmed entirely, or the ship inched across the glazed water slower than a person could walk. The great heat operated on the ocean's surface like a giant sauna, evaporating tons of water that returned to them in torrents every few hours. The evening darkness, which dropped like a shroud, offered little relief. Their clothes and boots became covered in a vile blue mold that required fierce scrubbing to remove. The saturated air, complained assistant steward Joseph Matkin in his diary, "seems to take all the life & energy out of a man."[1]

Challenger recrossed the Equator on March 12, 1875, in a flooding rain, barely in motion. Fresh provisions were long gone; they were down to salted meat and the last of the water. Coal, too, required rationing. Captain Thomson rarely fired the steam engine now, relying on old-fashioned sail even in a dead calm. Every surface, above decks and below, was slippery with damp. They slept in their own sweat and grew thin, pale, and irritable. Suhm, whose health had been fragile most of their two years at sea, broke out in nasty sores. In a letter to his mother, he described his shipboard existence: "Nothing but sea, water and more water, no fresh food at all and infernal heat." Drifting east in search of winds to bring them north, the ship seemed to crawl "like a parasitic snail" over the lifeless sea.[2]

Though the map showed islands aplenty, the insipid breeze kept them out of sight of land during this miserable month. Guam, with its running streams and fruit trees, lay tantalizingly to the east, but the contrary wind would not bring them

1 Matkin (1992) 228.
2 Willemoes-Suhm (1877) 159–60.

to it. The sheer enormity of the Pacific was brought home to them. Here, in a mere corner of it, was an ocean without limit, seemingly independent of all land or the dream of land and utterly skeptical of human presence. Phosphorescent fish were among the few entertainments. A hatchetfish (*Sternoptyx diaphana*) brought up by Suhm after night had fallen "hung from the net like a shining star."[3]

One of the company at least maintained good spirits during their moribund transit. While the ship idled in its own refuse, John Murray spent hours in the small boat scooping microscopic plankton into jars. Back in the shipboard laboratory, he waited for the evening's deposit from the dredge. But a surprise was in store. One morning, with Guam and the Mariana Islands beyond the northeast horizon, the content of the surface catch did not match the previous evening's deposit. The surface animals, in their millions, had died then disappeared. The abyss had swallowed them up leaving only a distinctive radiolarian ooze.

It was still dark on the morning of March 23 when the officer on watch ordered the boilers to be got up. Just after five o'clock, hands scurried aloft to perform the familiar ritual of shortening and furling sails, deemed useless for another day. Dawn revealed a sea as still as glass, a ghostly vapor rising from it. At six o'clock, *Challenger*'s steam engine belched into life. At ten o'clock, they paused for sounding. With 3,000 fathoms of line let out, it was assumed some mistake had been made, though the weights were still attached. They checked again at 4,575 fathoms—more than five miles deep—and this time the weights were found to have been loosed, indicating bottom. Hauling in the line, they brought up the remains of a thermometer pulverized by the great pressure and a sounding tube filled with red clay. The philosophers were intrigued but not convinced. In the afternoon, at as close to the same

3 Willemoes-Suhm (1873–1876) 24:81.

position on the featureless ocean as practical, the captain ordered soundings to be repeated. They attached a heavier weight and checked the line closely during its descent: 4,475 fathoms confirmed their record sounding, with two more thermometers smashed to powder.

Here, at the Challenger Deep, the carbonate-shelled foraminifera in their scummy chains of millions across the sea's surface had vanished from the bottom deposit. Instead of white *Globigerina* mud, the dredge delivered only lifeless red clay—ancient mineralized lava and pumice—threaded with microorganisms in their wild stellar variety. In these abyssal waters of the Pacific—five miles down—the radiolaria under the microscope amazed John Murray with their tenacious presence in the cold ooze. For the first time in their yearslong journey, after drawing hundreds of samples from the ocean floor, the silica-based protists preponderated on the seabed.

A new ocean chemistry was called for to explain the *Globigerina*'s disappearance. The radiolaria's tiny, lattice-shell skeletons and thorny spikes, made of silica, survived whatever chemical agent it was that dissolved their brother carbonate-shelled plankton in their journey from surface to floor. There was, moreover, a distinct cutoff depth. Above 2,500 fathoms, or thereabouts, the poster child of the foraminifera—*Globigerina*—filled the ocean floor. At depths lower, as in the tropical Pacific, the *Globigerina* ooze vanished.

Water sampling bottles attached to the sounding rod provided Murray with a vital clue: carbonic acid levels increased with depth. The professor's old dredging companion William Carpenter had likened the ocean to a vast set of human lungs: the oxygen of the enveloping atmosphere aerated the waters, while the carbon waste of ever-busy marine life breathed upwards from below. Animal activity was concentrated at the surface and at the bottom. The ocean floor, in particular, teemed with life and the accompanying rot of death, exuding vast volumes of carbon into the water layers immediately above.[4]

Below 2,500 fathoms, acidity levels reached a tipping point where the descending carbonate shells dissolved before ever finding their appointed grave in the ocean bed: the so-called calcite compensation depth. Thomson had long held to the conviction that the animals of the ooze lived and died in the abyss. The professor's error was forgivable: it was no small thing to expect these plankton, invisible to the eye, to freely float, eat, and reproduce in the sunny upper ocean while also possessing a bodily exterior robust enough to survive the miles-long descent to freezing depths. But survive they did—and Murray's radiolaria, it turned out, were the greatest trojans of all.

The concept of mobile tectonic plates was undreamt of in the Challengers' philosophy. They had already confirmed the existence of the mid-Atlantic Ridge, which would, a century later, provide core evidence for seafloor spreading. Then at the Challenger Deep they stumbled again across a dramatic expression of tectonics. Where the western edge of the Pacific plate sinks beneath the edge of its sister plate to the west—120 miles from the Mariana Islands—lies the Mariana Trench, the deepest undersea canyon on Earth. Near its southernmost tip is its deepest point, no more than a mile wide, where the ocean floor sinks a full mile deeper beneath sea level than Mount Everest rises above it.

HMS *Challenger* was not supposed to be in the vicinity of the Mariana Islands at all in March 1875. Light winds and a dwindling coal supply had frustrated their course since the Philippines. This fact, compounded by the tiny area of sea surface corresponding to the record deep, meant their chances against sounding along the Mariana Trench in that precise location were astronomical. But still the miracle occurred, with heroic implications for the expedition. For all the thousands

4 Carpenter (1869–1870) 480.

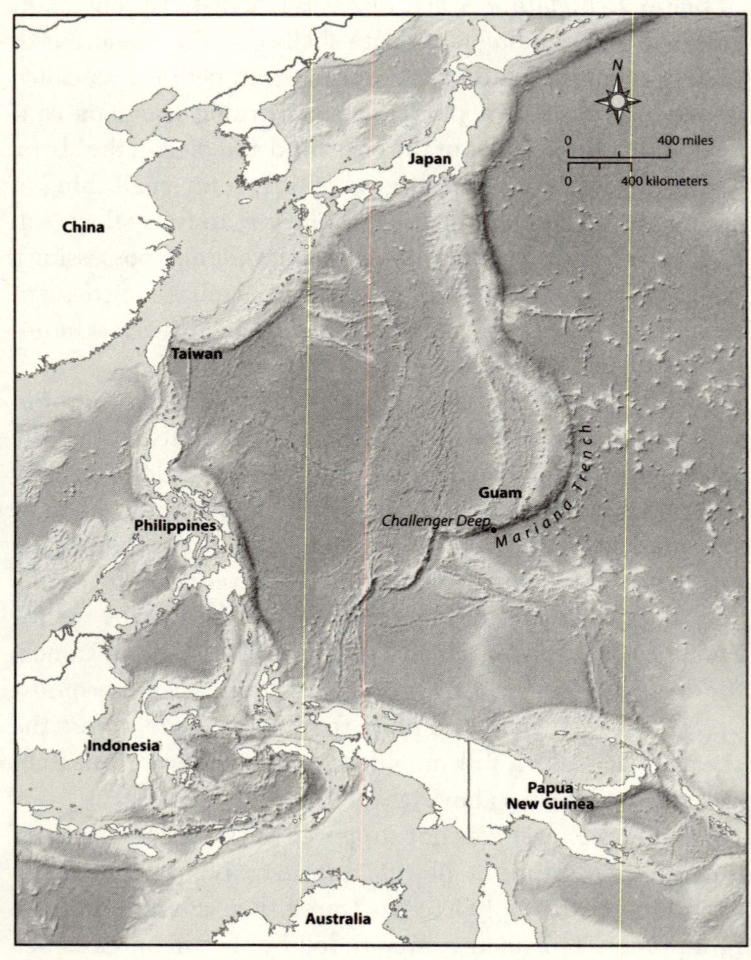

FIGURE 9.2 The Mariana Trench, deepest on Earth, with the Challenger Deep shown. Courtesy of wallacne / Wikimedia Commons (CC BY-SA 3.0).

of new marine species discovered by the expedition, and breakthrough insights like the calcite compensation depth, the *Challenger* name is best remembered in the popular imagination for this one fluky feat—their discovery of the deepest place on Earth, the Challenger Deep.

But this signature achievement is, in another sense, truly apt. *Challenger*'s research object was not the oceans in the abstract but specifically the *deep* sea and its mysteries—all that was least known about the life aquatic. On their return to England in the spring of 1876, Director Thomson promised full publication of the expedition's results within five years. By 1880, however, not a single volume had appeared. Questions were raised in parliament and the press by those who did not appreciate the complexity of the task: almost 6,000 *Challenger* specimens in glass bottles, tubes, and jars had been sent all over Europe and to the United States, a process which itself took several years. The writing of reports was then in the hands of specialists whose work schedules and dedication to the task Thomson could not control and with whom he was in infrequent contact. Correspondence on a single question or problem might take months. The inclusion of hundreds of illustrated plates—a spectacular visual legacy of the expedition—only added to the time and expense.

By 1878, early reports had begun to trickle in from experts in London, Edinburgh, and Berlin. But because submission deadlines were nonexistent, their subject matter was wildly variegated, even hodgepodge: essays on shellfish (the brachiopods), sea pens, and so-called seed shrimp, arrived in the first batch on Thomson's desk alongside analyses of whale bones and a green turtle fetus. Given the public's long wait for *Challenger* reports, and to avoid a sense of anticlimax in the debut volume's publication, Thomson decided to include a preliminary overview of the expedition's results. As a gifted systems thinker in the Victorian style and partisan of the deep sea, he was ideally suited to the task of an expedition manifesto.

Thomson's "General Introduction" to the first volume of the expedition's *Scientific Results* (1880) is a fascinating document because it distills what, to *Challenger*'s scientific chief, their principal aims and achievements had been. His focus on abyssal ocean conditions and the long-term evolution of their fauna is therefore telling. For a Victorian scientist with a necessarily sketchy understanding of geological time and a total ignorance of tectonics, the "General Introduction" offers a remarkably lucid, substantially accurate theory of the deep sea.

Thomson's key categorical distinction in the essay is between the familiar fauna of the seashore and the alien forms of the abyss: the glass sponges, crinoids, blind crustacea, and "grotesque" fish species found worldwide by *Challenger*. The uniformity of deep-sea life was the expedition's principal finding, which Thomson attributes to the work of polar bottom currents issuing from the north and south, thus creating a stable temperature environment for marine life. "Temperature," as Thomson states in his second general principle, "is the supreme condition which determines the distribution of marine animals." Modern marine biology is founded on this understanding. Temperature drives the push and pull of life under the sea. And on geological timescales, deep-sea communities are closely wedded to changes in climate—both incremental and catastrophically sudden.[5]

Thomson's stated reasons for the differences between shallow and deep-sea fauna were more speculative and fodder for the ongoing bitter controversy over evolutionary theory two decades after Darwin's publication of *The Origin of Species*. Thomson surmised, correctly, that marine life had originally migrated from the shallows to the deep sea. He attributed the present differences between coastal and deep-sea life forms to the constant changes in geomorphology and sea level along

5 Thomson (1880a) 48.

the continental shelves, in contrast with a stable deep sea (Thomson's most profound error in light of modern tectonic theory).

So far, uncontroversial. For his essay's conclusion, however, Thomson finally committed himself to the fray as a Darwin skeptic. The affinity of the abyssal creatures collected by *Challenger*—the sea lilies, glass sponges, etc.—with fossil relics of the ocean past offered "powerful support of the doctrine of Evolution," he declared. But he could find no evidence for natural selection as the mechanism for biological change. Among *Challenger*'s vast deep-sea collections, he saw no "insensible gradations" from one species into another, no "transition forms." "There is," he concluded, "no difficulty in telling what a thing is." Where Darwin looked for difference, Wyville Thomson found reassuring similarity.[6]

Thomson's views were not a throwback to species fixity, like Louis Agassiz's. But neither could he support the idea of sudden convulsions—of mass extinctions. Only the slow pulse of heat, now steady and strong, now faint and irregular, had ushered species to the margins of their range or over into oblivion, while reshaping others for survival in the new climate regime. Species changed, even went extinct, but not through competition. Rather, they passed through a variety of modifications in response to their changing environment until their powers of self-invention were finally exhausted.

Thomson's final throwaway line—"no difficulty in telling what a thing is"—provoked a rare public attack from Darwin himself. From its outset, the *Challenger* expedition had been viewed by the Victorian scientific community as an opportunity to prove or disprove Darwinian theory once and for all. If evolution were true, the deep sea would deliver a trove of living fossils visibly connecting ancient and modern life forms. Now, in his first public statement on *Challenger*'s findings, its

6 Thomson (1880a).

scientific director denied any such proofs had been found. Deep-sea fauna was the product of "progressive change" over eons, Thomson conceded, but this "slow and continuous" process was due to "slightly varying circumstances" in the ocean environment of whose laws "we have not as yet the remotest knowledge."[7]

For Darwin, the potential harm of Thomson's arguments to public confidence in evolution was serious enough for him to consult immediately with his most trusted advocate at the Royal Society, Thomas Huxley. As their first attempt at damage control, Huxley penned a careful review of the debut *Challenger* volume in the progressive journal *Nature*. He first lauded Thomson and his scientific staff as "the best equipped voyagers who ever left the shores of England for the purpose of enlarging the bounds of natural knowledge," and generously excused their delays in publication. On the lightning-rod issue of natural selection, however, he both criticized Thomson's opinion as "hardly so cogent as might be desirable," while also arguing that his main point—transitional forms are rare and unusual—was in fact perfectly consistent with Darwinian theory.[8]

Darwin was not satisfied with this polite exchange of views. The day after his friend's review appeared, he enclosed in a letter to Huxley an acid-toned rebuttal with directions to edit and forward it to the editor at *Nature*. He was both enraged at Thomson for his waffling on natural selection and worried that any public expression of anger would undermine his hard-won aura of Olympian remove from public debate over evolution. "If my manuscript appears . . . too contemptuous, too spiteful, or too anything," he wrote anxiously to Huxley, "I earnestly beseech you to throw it into the fire."[9]

7 Thomson (1880a) 49–50.
8 Huxley (1880) 1–2.
9 Darwin (1903) 2:388; see also Huxley (1900) 2:15.

In the letter rushed into print the following week in *Nature,* Darwin mocks Thomson for "not understand[ing] the principle of natural selection" and insists on his own full recognition of "the direct action of external conditions on organisms." The most heated section of the draft letter—in which Darwin excoriates Thomson as "a man who talks about what he does not in the least understand"—Huxley quietly deleted from the published version. Darwin's son and editor, Francis Darwin, still considered this the strongest language Darwin ever allowed himself in print—a measure of the perceived high stakes of the Challenger *Reports* to the public reputation of Darwin's theory.[10]

The Darwin-Thomson quarrel in the pages of *Nature* in 1880 might appear a footnote to the epic battle over evolution were it not for the fact of the decades-long "eclipse" of Darwinism in the wake of the Challenger *Reports*, and that the basic terms of their disagreement have been recently revived. Modern evolutionists have utilized Lewis Carroll–inspired figures of the Red Queen and Court Jester to distinguish between two rival theories of species change: the first model is classically Darwinian and emphasizes biotic factors, notably competition (Red Queen), while the alternate model invokes environmental perturbation, such as climate change, as evolution's principal engine (Court Jester). Viewed through this updated theoretical lens, Thomson's "General Introduction" to the Challenger *Reports* stands as an early contribution to the Court Jester school of Darwinism. Evolution as a working theory has been rendered infinitely more complex (and useful) with the advent of twentieth-century genetics. Geology, in turn, was

10 Darwin (1880) 32. On *Challenger*'s return, Henry Moseley warned Darwin by letter of Wyville Thomson's doubts. In his reply, dated November 22, 1876, Darwin took the larger view: "I am not disappointed at what Wyville Thomson says: as long as a man believes in evolution, biology will progress and it signifies comparatively little whether he admits natural selection and thus gains some light on the method, or remains in utter darkness."

entirely upended by the new tectonics paradigm in the 1960s. Notwithstanding these intervening scientific revolutions, the tension between the basic positions outlined in the 1880 Darwin-Thomson debate—biotic versus environmental—persists as an original fault line in evolutionary thought.[11]

The Challenger Deep is a site of modern myth and has attracted both Jules Verne–style adventurers and cutting-edge scientific research. The first human-piloted submersible reached the bottom of the abyss in 1960, followed a half century later by film director James Cameron of *Titanic* and *Avatar* fame, and more recently by a Chinese research craft.

Remoteness is undoubtedly the lure for these pioneers, though that feature is liable to overstatement. The unique topography of the deep-sea trench—where the hard edge of an abyssal plain plunges through the soft sediment of a continental plate—creates a funnel effect. Despite the gentle gradient of their slopes—rarely more than 10 degrees—any and everything from the ocean's surface will gravitate to the trenches of the deep sea, including the carcasses of fish and whales, and plant debris deposited by river outlets from faraway coasts.

A unique, extreme-deep-sea community has evolved to feast on this bounty, including a cockroach-sized crustacean (the amphipod *Hirondella gigas*) able to digest nutrients from wood particles. Some fraction of the floating forest the Challengers encountered off the coast of New Guinea in February 1875 surely made its way at last to the lightless floor of the Challenger Deep. But now, so do microplastics. In excess of a million metric tons of plastic waste issue annually from the rivers of New Guinea alone, itself a tiny fraction of the overall global output of 275 million metric tons in 2010, a total expected to rise by orders of magnitude in coming decades. The deep-sea trenches of the world, including the Mariana, are

11 For Darwinism's "eclipse," see Bowler (1983) and Benton (2009).

destined to "be the ultimate sink for a significant proportion of the[se] microplastics."[12]

The tiny stomachs of *H. gigas* recently sampled from the Challenger Deep contained an unappetizing stew of human waste products: synthetic fibers from machine-washed clothes, cow DNA, and fifty times more toxic polychlorinated biphenyls (PCBs) than sampled from crabs native to the most polluted river in China. The human-piloted submersible *Fendouzhe*, exploring the Challenger Deep in 2020, discovered a unique biome of starfish, sea anemones, and mollusks inhabiting the Earth's most remote ocean floor, proof of the tenacity of marine life in extreme conditions. They also found plastic shopping bags, a television cable, and a beer can.

12 Peng et al. (2020) 1.

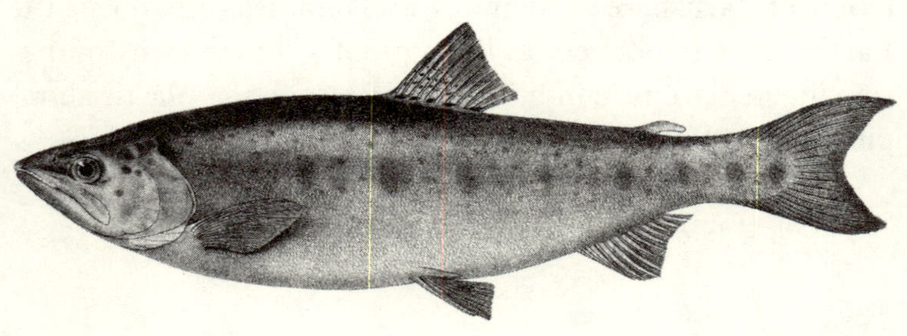

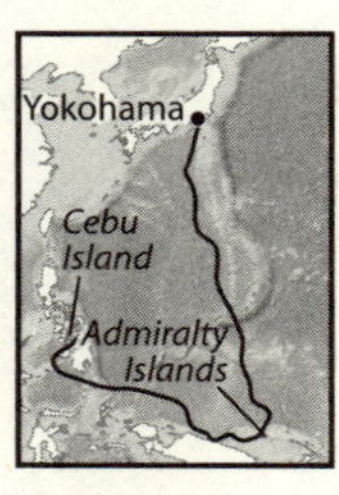

FIGURE 10.1 The cherry salmon (*Oncorhynchus masou*), indigenous to Japan. *Scientific Results: Zoology*, vol. 1, plate 31 (1880). (Inset) HMS *Challenger* Leg 9. Cebu Island to Yokohama, Japan. January–April 1875.

CHAPTER 10

A Salmon's Pilgrimage

Yokohama, Japan
32°26′ N, 139°38′ E

The cherry salmon (*Oncorhynchus masou*) is the only Pacific salmon species exclusive to Asian waters. Various theories attach to the origins of its common name in Japanese—*sakuramasu*. It might be the pink markings the adult fish adopt in spawning season, which match the colors of the blossoming cherry tree (*sakura*), or that cherry trees are in bloom when the young masu salmon embark on their downstream migration to the sea.

The cherry salmon's odyssey from river to ocean and back begins with a discreet underwater explosion: the hatching of millions of tiny eggs in streams deep in the forested mountain glens of northern Japan. Spurred by springtime warmth, the maturing masu eggs—orange with crimson spots—shudder and split in their gravelly nests (called redds). How rapidly the unborn masu develop, if they survive at all, depends on the precise temperature of the water bathing the eggs. If all is well, sometime in May or early June, the masu fry emerge in teeming broods, each the length of a finger joint.

The first notable event in the young salmon's life is the death of its mother. She protects her offspring for as long as she can, a week at most, before her swimming power suddenly ebbs and the current bears her body away. Left to fend for themselves, the masu fry flock together, finding shelter in shallow waters near their birthplace.

With growth comes confidence to disperse but also a keener sense of threats from predatory fish and birds. The downstream end of a deep pool offers the best sanctuary in the rushing stream. There the little cherry salmon fights to maintain its hold in the current, darting out to feed on the larvae of midges and insects fallen in the water. At night, as visibility dims, it retreats to the pool's edge or behind rocks where the current is calmer. The masu are social creatures and quickly establish a working hierarchy. Bigger, more successful feeders dominate the school.

This original golden summer cannot last, of course. In September when the temperature falls, the juvenile masu, now officially a smolt, sets out on the first phase of its lifelong migratory adventure, seeking out warmer waters downstream on the plain near the coast. It finds a natural haven along the river banks, with bamboo shards and dead grasses drifting above. Once established, the cherry salmon grows in earnest to 5 inches (12 centimeters). Its flanks turn silver-white, the colors of its first winter.

At the return of spring, the next generation masu fry are just emerging from the pebbly bottom upstream when their older siblings, lurking miles downstream, make their bid for the open ocean. In an epic circular journey lasting two years, the juvenile masu will follow the seasonal warm waters north to the Sea of Okhotsk then south to the Sea of Japan, or east via the Tsugaru Strait to the Pacific Ocean. En route, it will snack on rich oceanic offerings—mackerel, anchovies, and a full plankton buffet—growing in size until it reaches a mature adult length of 20 inches (50 centimeters). Most adult cherry salmon are female, and her life mission is essentially nostalgic: a return to her natal river mouth followed by an arduous pilgrimage upstream to spawn.

Unless the masu's ocean odyssey never happens at all. The cherry salmon that the *Challenger* naturalists purchased at the Yokohama fish market in April 1875, for example, measured a mere 11 inches. This juvenile masu—described as silvery with spots but otherwise "eviscerated" by its experience—had most likely graduated to an ocean existence only a few days before strangulation in a net lowered from one of the hundreds of single-masted fishing boats *Challenger* sailed among on the sheltered waters of Tokyo Bay.[1]

The Challengers were not sure, based on their damaged specimen, whether they had chanced upon a salmon entirely new to Western science or merely an individual of the same species recorded by American Matthew Perry's expedition two decades prior—as part of a naval campaign of intimidation that initiated the historic "opening" of Japan to the West.

Either way, the irony of an Asian salmon fished for millennia in Japan earning the distinction of "new species" by HMS *Challenger* symbolizes the near total lack of cultural contact between Europe and Japan prior to the Meiji Restoration of 1868. Arriving at the outset of that seismic realignment in Japanese foreign relations, the Challengers, like many Western visitors of the time, found themselves torn between a touristic sense of enchantment at the feudal culture still everywhere in evidence and their colonialist desire for an open, commercial Japan managed as a European client state.

From the Japanese point of view, the shock of American gunboats in Tokyo Bay in 1853 had seeded a generational commitment to reclaim national independence through adoption of Western technologies. Yokohama, a small fishing village, was chosen as the site for this grand international experiment. By the time of *Challenger*'s visit, the streets by the port had already taken on the hybrid, global character of modern Japan. Traditional, elegant shops selling Japanese silk, tea, and interior décor stood alongside colonial-style consulates, a railway, and a naval academy run on European lines.

1 Günther (1880) 1:72.

The Japanese proved themselves active sellers *and* buyers in this emerging global emporium. The Meiji-era transfer of European technology to Japan included fishing vessels and equipment whose enthusiastic adoption—just underway at the time of *Challenger*'s visit—would have deep consequences for the global oceans of the twentieth century and beyond, not to mention the unique lifeways of *Oncorhynchus masou*.

A welcome breeze, sufficient to churn a little foam of surf at the bow, delivered HMS *Challenger* to her berth in Tokyo Bay amid a flotilla of British and American warships. After two and a half years at sea, the ship—beset with barnacles and smelling of rot—was overdue for dry-dock repairs. The Challengers, meanwhile, were eager for firsthand experience of a country and people they knew only from picture books and, more recently, heady newspaper accounts of revolution.

Tokyo Bay was, in fact, a giant harbor full of little, forested islands and tidy coves and bays, populated by square-rigged fishing junks built according to a single mandated design. The recently deposed military shogunate, which ruled Japan from the early 1600s, had outlawed all oceangoing vessels to deter foreign contact. The junk's conspicuous open stern ensured that no captain was tempted to defy the order and that the Japanese fishery confined itself to the coasts.

With *Challenger* anchored in the bay to sound and dredge, John Murray took the small boat across to greet members of the nearby fishing fleet. Given that he took no translator with him, it's unknown how he contrived to discuss fishing practices with the locals. No doubt the imperial remit to go and do as he pleased—in the spirit of Commodore Perry—played its part in his polite reception.

On board a local fishing boat, Murray observed several neat adjustments to government standards. A bamboo frame, from head to stern, made all watertight and supported movable sheets of straw for warmth. The fishermen themselves wore rain jackets made of what appeared to be paper and tended a little charcoal fire amidships. One crew member beat a wooden rattle—to charm the fish—while the rest hauled in a maze of nets and lines. The deck was soon littered with a spectacular bounty. On the hooks, strung six feet apart on their lines, Murray noted sea pens, several starfish, a bamboo coral, and a variety of fish entirely strange to him. As at Cebu, the Challengers would rely on the locals' dredging technology to pluck the sought-after *Hyalonema*—the glass rope sponge—from its rooted place in the silty bottom.

Back on board *Challenger*, Murray found his fellow philosophers gathered around a black shark with green eyes and a giant crab the size of a child just disgorged from the trawling net. Its spindly legs were twice the length of its back, which was encrusted with barnacles. The amazing *Macrocheira kaempferi*—the Japanese spider crab—was known to Western naturalists. Philipp Franz von Siebold—the only European naturalist granted roaming rights in Japan prior to the Meiji Restoration—had immortalized *M. kaempferi* in his 1830s studies of Japanese flora and fauna, which introduced the hydrangea, among other novel beauties, to Europeans. Siebold's

FIGURE 10.2 Japanese spider crab (*Macrocheira kaempferi*). From *Scientific Results: Zoology*, vol. 1, plate 31 (1880).

subsequent career in Japan, as botanist-diplomat and *agent provocateur*, blurred the lines between scientific research and imperial politics. That *Challenger* herself was mostly indistinguishable from the warships at anchor in Tokyo Bay bore out the same reality.

Deeper into the harbor, *Challenger* sailed past so-called Treaty Point where, in March 1854, Commodore Perry had made Japanese officials an offer they couldn't refuse. Those historic concessions on international trade gave birth to the booming port city now coming into view. More eye-catching than Yokohama itself, at first, was the blue-tinted mountain range beyond it. Great swaths of forest covered the mountainsides with snow-capped Mount Fuji the crowning glory, "looking like frosted silver." The scene was already a popular postcard view.[2]

The fish market at Yokohama, by contrast, was like nothing they had ever seen. The Challengers wandered among stalls crowded with dripping oysters, staring pink salmon on slabs, turtles weighing 300 pounds, and half-ton porpoises hung from hooks. They recognized a ribbonfish (*Trachiurus lepturus*), staple of the modern Japanese fishery, and the ubiquitous John Dory (*Zeus faber*), now under threat globally from overfishing. They inspected a red seabream (*Pagrus major*) nicknamed the King of a Hundred Fishes, a showpiece dish prized in Japan as a bringer of good fortune; a poisonous puffer fish (*Takifugu rubripes*) valued for medicinal powers but now also endangered; a mackerel (*Trachurus japonicus*) later served to the Challengers as sashimi; and the distinctive Luna lionfish with its cultured mane (*Pterois lunulata*), soon to escape Asian waters to plague coral reef systems worldwide.

With stewards from the *Challenger* in tow, the philosophers paid Mexican silver dollars for a white-spotted octopus, sword-tip squid, a tortoise, an eagle-winged stingray (to be cooked

2 Spry (1878) 274.

how?), as well as sea cucumbers, groupers, flatheads, sole, rockfish, perch, grubfish, and a lone cherry-speckled salmon possibly new to science—*O. masou*. They haggled alongside housemaids for the choice specimens then lugged their bags, slippery and reeking, back to the ship.

An obvious question emerges from these accounts. How did such a cornucopia of fresh fish come to the Yokohama market in an era before diesel-powered ships and refrigerated trucking? Much has been made of the so-called miracle of Japan's rapid modernization in the aftermath of the Meiji Restoration. It could not have occurred, however, without the deep, proto-industrial appropriation of natural resources already in place. Advanced technologies of capture and distribution were conspicuous in Meiji Japan's fishing industry, which first developed—out of sight and mind of Europeans—in the storied Inland Sea.

The Challengers versed in travel literature knew the Inland Sea as the the "Mediterranean of the Orient," where modern steamers en route from Osaka to Nagasaki wove among "an ever moving mass of white sails." Fishing junks in their hundreds bobbed on waters still as glass among the wooded islands.[3]

At the sight of warlike HMS *Challenger*, the Japanese fishermen of the Inland Sea stopped work and stared in amazement. In two years at sea, the Challengers had rarely enjoyed so rapt an audience for their trawling operations. One boat boldly pulled alongside with a gift of sea worms called penis fish (*Urechis unicinctus*), presumably for bait. Viewed from the main channel of the Inland Sea, the coast reminded Willemoes-Suhm of the celebrated forest vistas of the Rhine. But closer to shore, they saw that each cove and inlet had its populous fishing village, with waves of yellow grain stretching behind and boats in constant traffic.

3 Spry (1878) 299.

Before the rise of the shoguns, the Inland Sea had been a pirate stronghold. The sight of an abandoned castle on a hill was a poignant reminder that the tides of history had, in turn, swept away the shogun and his allies. The local lord (*daimyō*) had recently been deposed and his samurai retinue disbanded. Forced from his ancient fort, the lord now subsisted on a government pension in a modest house in town. Among the multitude of crustaceans swarming out of *Challenger*'s nets in the Inland Sea was the so-called samurai crab. Its patterned carapace, looking like the face of an angry warrior, seemed to project an added note of shocked despair.

Sweeping political changes in Meiji Japan inevitably brought upheaval to the fishing industry, which had evolved over thousands of years since the ancient Jomon peoples of the archipelago first built stone weirs to trap salmon and wove nets to trawl for herring. The forced abdication of the *daimyō* had dissolved traditional feudal control over both land and sea, instantly transforming the seafood economy. Bountiful tributes of fish once owed to the lord and his entourage could now go straight to market.

The year of *Challenger*'s visit, the new government furthered their policy of fisheries reform by abolishing the ancient offshore fishing zones preserved for local villages. Their goal was to attract newcomers and expand the marine economy as a whole. To the same end, the long-standing monopoly of fish wholesalers—who sold equipment to cash-poor fishermen on credit while profiting from their catch—was broken. Labor and capital flowed seaward. In short, the fishing industry in Japan in the spring of 1875 was on the brink of enormous growth, with a corresponding increase in impacts on marine life. For example, herring populations—a Japanese staple—would face steep decline by the end of the century.

Meiji-era growth in Japan's fishing industry was on the back of an already technologically advanced, bustling trade, as the Challengers witnessed firsthand. The ubiquitous junks de-

FIGURE 10.3 Hokusai Katsushika, *Line Fishing in the Miyato River* (1833). Hokusai's images of Japanese landscape and culture, centered on its waterways, belong to the wave of Japanese cultural imports to the West beginning in the 1870s. Hokusai / Oceans of Wisdom Series. 18.5 × 25.5 cm. Woodblock print ink and color on paper. Minneapolis Institute of Art.

ployed miles-long lines to fish for the lucrative seabream and sharks and rays for the China market. Meanwhile, self-employed fishermen in small boats used handlines for catching sea bass, cod, and young bream during the day, and mackerel and squid at night. Their torches lighting up the bays of the Inland Sea looked like fallen stars. Night fishing also included nets, some of them industrial in size. A 300-fathom net, deployed by four boats, encircled schools of mackerel, while the fishermen yelled and waved torches to confuse their prey.

An 1870s government report on the fishing trade in Fukuoka on the Japanese west coast details the dizzying range of net technologies already available to Meiji-era fishermen. Gillnets—hung vertically like a curtain from a rod—trapped black porgy in the winter and bass, cod, and flathead in the spring. Other gillnets were designed specifically for yellowtail and shad.

Netting techniques varied, species by species. For halfbeak and bass, the fishermen cast their gillnet in a semicircle facing the shore then rowed the boat to the landward side to drive the frightened fish into the trap. Herring fishing, meanwhile, required a full village effort. Two lookouts on a hill scanned the sea for telltale red streaks below the surface, while thirty men on boats deployed a seine net in a long arc from the shallows to the shore. On the beach, dozens more confederates—men, women, and children—waited to haul in the net full of splashing sardines.[4]

Fishing was seasonal, and *Challenger*'s visit coincided with specific fish migrations and corresponding tactics for their capture. On the fishing grounds of the Inland Sea, they observed handlines cast for squid, longlines for rays, nets for bream and yellowtail, and dredges for scallops and sea cucumbers. At night, the boats set driftnets for flying fish. Women patrolled the beachside shallows for shellfish and the springtime seaweed called *okyūto* (in summer, they dived for abalone). From wooden piers, fishermen lowered pots on the seabed as traps for octopus.

For generations, seafood has provided half of all protein in the Japanese diet, the highest ratio in the world. In the preindustrial era, only a fraction of this mighty, year-round smorgasbord could be sold and consumed fresh—much of the herring catch was converted to fertilizer. Techniques for drying and salting fish had been perfected over centuries, delighting even the Challengers' suspicious palettes. The sophistication of their hosts' fishing and preservation technologies was cause for wonder, as was the sheer bounty of the trade. Victorian-era visitors to the coastal villages of Japan reported being repelled by "the all-pervading stench of dried fish and seaweed." But this collective Japanese dedication to the sea—and centuries

4 See Kalland (1995) 99–115.

of inherited knowledge—would help set the stage for their supremacy in the global fishery in the decades to come.[5]

Where the warm, northward-flowing Black Current of Japan's Pacific coast meets the colder waters of the Sea of Okhotsk lie some of the most productive fishing grounds in the world. For millennia, Japan's export fishery had been directed from there westward into the Sea of Japan, to neighbors China and Korea. The vast Pacific to the east lay unmapped and unapproachable by Japan's lightly built fishing fleet. Project *Nanshin*—post-Meiji oceanic expansion—began humbly with an 1860s policy to apprentice Japanese crews on American whalers. The Meiji government then sent officials further abroad, to Europe and the United States, to learn the latest advances in Western fish processing and aquaculture. By decade's end, Japan had piloted its first salmon hatchery for the artificial reproduction of *O. masou*.

The year 1875 also saw a treaty between Russia and Japan that opened the teeming, cold salmon runs of Sakhalin Island to Japanese fishers, rights that later extended north to Kamchatka. At home, the institutionalization of fisheries management gathered pace with the establishment of a government-backed Fisheries Society and a professional institute for fisheries science. The *Nanshin* imperative was boldly framed in a statement from the Fisheries Society in 1884: "For our country to overtake England and America . . . there is no other foundation apart from developing the fishing industry."[6]

In 1908, European-style trawlers were adopted for high-seas fishing. Canning technology, imported from France, was another decisive innovation. A Yokohama-based firm first began selling canned sardines on the British and American markets in the 1880s, followed by canned crab and salmon from the Sea of Okhotsk. By the 1930s, when Japan controlled

5 Blakiston (1883) 5.
6 Quoted in Muscolino (2013) 59.

a full third of the global fishery, it was selling millions of cases of canned fish abroad annually in return for vital commodities: rubber, iron ore, and oil.

Japanese market domination was enabled by a flotilla of so-called factory ships. These patrolled the Pacific Ocean all the way to the Alaskan coast, south to the Sunda Strait and Australia, and west to the Indian Ocean and the whaling grounds of Antarctica. Notorious for brutal working conditions, Japanese factory ships represented the acme of twentieth-century industrialization of the oceans. Hundreds of tons of fish could be processed on board a single vessel operating around the clock. When flash freezing technology advanced after World War II, capacity increased exponentially and with it the global market for fish products. The number of fishing vessels in the Asian Pacific doubled again between 1976 and 2000. Global seafood consumption at the turn of the twenty-first century was seven times what it had been in 1950.

Alarm bells at a sudden worldwide decline in commercial fish stocks crescendoed in the 1970s, resulting in the United Nations–driven "enclosure" of the oceans. National fishing fleets were henceforth to be restricted to a 200-mile zone from their shores. This new geopolitical arrangement could not halt the collapse of pelagic fish populations, however. Probably one-third of commercial fish stocks are currently overfished, though the global trade "effectively masks the successive depletion of stocks." The decades since the crisis of the 1970s and 1980s have seen an increasing, sometimes desperate, reliance on seafood engineering—called aquaculture—to appease the human hunger for fish protein and all the mouth-watering delicacies of the sea.[7]

Back in the coastal villages of Japan—where arguably it all began—local fishing is a shadow of the trade the Challengers saw and admired. As early as the 1930s, officials were report-

7 Srinivasan et al. (2012) 544.

ing "the devastation of the coastal fisheries," even as Japanese high-seas trawlers swarmed the globe. By the 1970s, bream and other commercial fish populations in adjacent seas were crashing. The Japanese—who exported 70 percent of their salmon catch a century ago—now import nearly half their seafood, including salmon from distant Norway. The arc of ocean globalization—begun in the *Challenger* era—is complete.[8]

In busy Osaka, on the shores of the Inland Sea, Suhm went souvenir hunting with Henry Moseley. They wandered along streets that looked "like one enormous antique shop." He bought a wooden statue of the Buddha and an old manuscript with illustrations from Japanese myth, both for cheap. The shop owner told him ancient Japanese scrolls were no longer popular; his customers wanted only modern, technical volumes and books and maps in English. As for his friend Moseley, the Englishman went into a buying frenzy, taking with him "almost all of the beautiful pictures and manuscripts" before Suhm could reach for his purse.[9]

Ship's artist, Jean Jacques Wild, meanwhile, took himself to the craftware district. There, he was astonished at the quality of bronzes, porcelains, and carvings in wood and ivory for sale, which back home would be housed in a museum. To his European eye, the obvious superiority of Japanese design was "rather humiliating." Later, in Kobe, he experienced something like Stendhal syndrome in its Asian iteration. The sheer visual novelty of Japanese street life—the fashions, the houses, and "the thousand objects exhibited for sale"—all so tasteful, so decorative—made him feel he was "walking in a dream." Wild's is an early account of *Japonisme*—an aesthetic infatuation with Japanese culture that would reverse-colonize European arts and manufactures in the ensuing decades. His

8 Quoted in Tsutsui (2013) 36.
9 Willemoes-Suhm (1877) 170.

most uncanny experience was at an exhibition featuring a simulated English-style drawing room with furniture and fussy Victorian décor replicated in minute detail. The locals flocked to it and seemed vastly amused.[10]

Fascinated by what they saw at the Yokohama markets, temples, and tea houses, the Challengers were at the same time acutely aware of being assessed for what value they, as Western "barbarians," might bring to the new Japan. Added to these complicated sensations was abundant evidence that Japanese culture had thrived in its long independence from Europe. By subtle degrees during their two-month sojourn, they found themselves turned from colonizers to tourists.

To journey to the interior of Japan in the 1870s required complicated permissions and passports, a legacy of Tokugawa-era isolationism. John Murray was the most adventurous of the philosopher-tourists and set out on a week-long trek in the Nikko mountain range north of Tokyo. He was accompanied by thousands of local pilgrims to the great religious shrines bordering the shimmering alpine lakes. His journey likewise traced a path parallel to the great annual migration of cherry salmon upriver to their spawning grounds and is a unique record of that riverine landscape before its twentieth-century transformation.

A vivid account of Murray's Japan journey was left by his travelling companion, sublieutenant Lord Campbell, whose father, the Duke of Argyll, had recently served as secretary of state for India in the Gladstone government. Young, blond, and insouciant, *Challenger*'s Scottish lord cruised about the world with an ease he considered his birthright. Campbell was the most popular of the officers among the men, while the gentleman-naturalists actively sought his company—a case of title transcending professional rank (feudal holdovers were not unique to their Japanese hosts). Campbell did at least ex-

10 Wild (1878) 144, 150.

press embarrassment at being hauled up the Nikko mountain in a rickshaw (recently introduced as a source of employment for Japan's redundant warrior class).

On the highway to Nikko, rushing streams coursed through the middle of the mountain villages. Fishmongers on the roadside hawked a pungent variety of dried fish, from salmon to minnows, in great overflowing baskets. Murray and Campbell wondered at the colorful paper designs of fish hung from poles in front of the houses. They were told these signified a child born to that household that year: each fish a child. In the evening at the inn, they sat cross-legged on the floor of their furniture-less room as the servant girls brought plates of raw fish and vegetables. They afterward fell asleep to "the splashing drone of the river."[11]

In the morning, they learned the river had flooded, blocking their advance. So they whiled a day in contemplation of the gorgeous view from their tea house balcony: the town built of wood below them, the buzzing forest, and the silver-flecked stream rushing beneath. The air had a champagne freshness, as different from the *Challenger* lower deck as could be imagined. Ahead of them, the river shallowed and widened, "flowing rapidly over a stony bed" before narrowing again and vanishing through a cluster of cypress trees. In the distance loomed the bright, snow-capped Nikko mountains, their flanks tinted violet like in an Old Master painting. The Japanese country tea house (*chashitsu*)—Campbell confided wistfully to his journal—"stands quite alone among the pleasures of this world."[12]

Arriving at last at the foaming river, they found a sacred bridge. For centuries, the shogun had crossed here at the head of his colorful retinue, en route to pay homage at the Nikko temples. This tradition, like so many others, had been abruptly

11 Campbell (1877) 351–52.
12 Campbell (1877) 362.

discontinued after 1868. Their own pilgrimage led them close along the river. As the road grew steeper and rockier, the midstream of the water churned more violently. The quiet pools along the banks turned a glittering translucent green.

They were nearing now the summits of Nikko. At the crest of a ridge, Murray and Campbell looked across at a great waterfall, 700 feet high, pouring into a deep ravine blue with mist. They crossed carefully between the waterfall and cliff where they came across a suite of water gods sculpted in stone. After a short hike, they arrived at sparkling Lake Chūzenji, favored by the shoguns and, in decades to come, by European diplomats for their summer houses. It banks were lined with the pink blossoms of cherry trees in bloom.

The next day, as they began their return descent, they bought and grilled "some delicious little fresh trout" for a picnic by a stream. Given the season, this memorable lunch enjoyed by the *Challenger* naturalist and his aristocrat companion might well have been a plate of juvenile *sakuramasu*—otherwise known as cherry salmon.[13]

Murray and Campbell, absorbed in the pleasures of mountain tourism, had little attention to spare for (or instruments to measure) the mass springtime migration unfolding in the river before them. At this southern limit of their range, the adult *O. masou*—after a two-year ocean residence about which little is known—return to the coast in the springtime and reenter their birth river just as the snow begins to melt. They are the first of Japan's salmon species to make their homeward run. The larger males take the lead, with females and smaller-bodied males in the rear. For fishermen lying in wait on the Pacific coast of Honshu, this is the most productive season for catching cherry salmon.

13 Campbell (1877) 361.

The masu that manage to dodge the fatal gillnets offshore and enter the river swim first to downstream pools and quiet tributaries, as if girding themselves for the upstream battle to their spawning grounds. The journey will last several months. During this initial hiatus, *O. masou* undergoes the last of its metamorphoses: the male snout grows hookwise, and the fish's sides darken from silver-white to its signature patterned cherry-pink. This bodily maturation is the masu's cue to dive upstream. It travels mostly in the morning and at dusk. At first, it feeds as it goes. Once summer comes, however, during the last desperate race to its spawning grounds, *O. masou* stops feeding altogether, as if resigned to its sacrificial fate.

Izaak Walton, in *The Compleat Angler* (1653), was the first European to document the unique pilgrimage of salmon to their natal rivers. In feudal Japan, the heroic salmon was likewise celebrated in books and prints, though only the indigenous Ainu people of the north enjoyed *Oncorhynchus* as part of their regular diet. Elsewhere, salmon was considered a delicacy reserved for the nobility. In the eighteenth century, an imperial edict—*Tanegawa No Seido*—designated salmon spawning streams for conservation. Then, with the Meiji Restoration, the pink, chum, and masu salmons converted overnight to full-blown commercial fisheries.

In Japanese waters from the late 1870s to the early 1890s, the annual salmon catch averaged 7 million with a peak of 11 million in 1889—a rapacious, unsustainable harvest. The steady depletion of wild stocks has, for more than a century, driven the artificial reproduction of Japanese salmon in fish-farm hatcheries. Adult salmon, swimming upstream, are caught by the thousands and the females stripped for eggs. The so-called dry method of artificial insemination requires the eggs be pressed out from the female into a sterilized pan, then the sperm added, and the two mixed gently (before automation, this was done with a feather). Add water: instant fertilization. The resulting fry are raised in captivity before their mass release into the wild.

The first domesticated salmon production in Japan dates from 1879, modeled on a hatchery in Maine. A century later, 80 percent of salmon fry descending Japan's rivers into the sea were being reared artificially. Of the billion or so fry released annually, only 1–2 percent return upstream in subsequent springs as mature, spawning adults. It is one of the most drastic human interventions in any animal life cycle on Earth.

The Japanese hatchery system is economically profitable but ecologically dangerous in the longer term, for several reasons. The domestication of salmon selects for a less resilient fish. Hatchery-raised salmon suffer from undeveloped sensory organs (for lack of use), rendering them less able to identify prey or avoid predators. Salmon behaviors are likewise compromised by captive rearing. Keepers feed their hatchery fry from above, training them to populate surface waters that, once in real-world rivers, exposes them to threats that bottom-dwelling wild salmon naturally avoid.

Domesticated salmon are also more aggressive than wild salmon, pitting the two populations in unhealthy competition as well as—once again—increasing their vulnerability to predation. Where a wild salmon will lurk behind a well-placed rock, a hatchery salmon will blunder into the open. And because the hatchery system does not filter for inbreeding, the *Oncorhynchus* population overall is less genetically diverse, making it less resilient to viruses and environmental stressors such as climate change.

Indications are that warmer ocean waters are already affecting salmon maturation rates and body size. Hotter temperatures alter zooplankton populations, impoverishing the salmon diet, while adult fish in overheated coastal waters will not linger long enough to fully mature. The resulting smaller female salmon is less fertile, further driving down natural birthrates. This is on top of decades of overfishing with nets that disproportionally trap larger salmon, a case of human industry driving evolution toward diminished size and fitness.

In short, your average hatchery salmon is reckless, insensitive, and dumb. Its success depends entirely on overwhelming numbers. With the annual release of a billion hatchery fry into Japan's rivers, human-engineered salmon are driving wild salmon toward extinction and lowering the resilience of the salmon community overall. This Anthropocene salmon population is less adapted, less diverse, and more prone to collapse.

Massive hydro-engineering projects dating from the 1960s—built to stem chronic flooding problems in Japan's agricultural zones—have likewise played a major role in historic salmon stress. Concrete dams, which replaced the natural waterfalls marvelled at by John Murray and Lord Campbell in 1875, prevent upstream salmon from reaching their spawning grounds. In the last sixty years, fragmented riverine habitats have extinguished wild salmon populations across dozens of Japanese rivers and streams. Dams also increase water temperatures while altering the composition of river sediment, making it more compact and uniform. Salmon need loose gravel river bottoms to nest in and rocks and boulders to shelter behind. Compacted sediment is less organically rich, effectively starving growing salmon of food.

Of the native Japanese salmon species, *O. masou* has suffered worst from the combined effects of aquaculture and dam construction. The young *sakuramasu* remain in rivers a full year longer than other salmon and are at greater threat from freshwater pollution and engineering. Despite improvement of hatchery techniques that have inflated the populations of pink and chum salmon to historic highs—the current salmon catch in the North Pacific is about 500 million fish annually—*O. masou* have declined sharply since the 1980s and are now officially endangered. The masu fishery that harvested more than 2,000 tons in the 1960s has declined to fewer than 500 tons—this despite the annual release of 10 million fry from fish farms.

Ironically, the failure of *O. masou* hatcheries means that the proportion of natural cherry salmon in the wild is greater than for other salmon, even if overall numbers are dwindling. With recent improvements in river water quality and the dismantling of some dam infrastructure, the outlook for cherry salmon is brightening in some respects. But just as death cancels all appointments, the ever-mounting global market demand for salmon—and climate change—may yet render all conservation measures moot.

For average Victorians of the *Challenger* generation, the idea of humans ever depleting the oceans of a keystone fish was unimaginable. Murray and Campbell enjoyed their lunch of grilled *sakuramasu* by an idyllic (and undammed) mountain stream without a qualm. What irony, then, that the first signs of a shrinking salmon population in Japan lay only a decade in the future. The advent of steam trawling and the transfer of fishing technology from the West to Japan brokered the decline—itself a harbinger of the now critical pressures of overfishing on global marine life.

Honored with a farewell audience with the Japanese emperor, Wyville Thomson chose Lieutenant Campbell to accompany him to the royal palace in Tokyo. The Scottish lord's addition to the *Challenger* officer corps was designed for just such occasions. The British ambassador, Sir Henry Parkes, led them into the Mikado's presence, whereupon he personally delivered a letter from Queen Victoria. Campbell records that the Japanese officials present were all resplendent in European formal dress, capped by Emperor Meiji himself in an impressive "swallow-tailed coat laden with gold embroidery." The restored sovereign neither spoke nor smiled and left it to an attendant to express his pleasure at *Challenger*'s visit to Japanese shores. It was a meeting marked by mutual curiosity and incomprehension in equal measure.[14]

The ambassador and Lady Parkes accompanied the party back on board the *Challenger* where, for the first time, they saw the homeward pennant flying from the mast. The entire Pacific Ocean remained to be dredged and sounded, treacherous Cape Horn to navigate, and the Atlantic to cross one last time—but sight of the long, blue pennant lifted the company's spirits. The thought of home was some recompense, too, for leaving the attractions of Japan behind. Their watery pilgrimage, like the salmon's, was yet incomplete and fraught with dangers.

Tokyo Bay, crowded with fishing boats and warships flying the Union Jack and the Stars and Stripes, required careful navigation. But when at dusk they entered the powerful *Kuroshio*—the Black Current—flowing east from the coast toward the soft light of the Pacific, the new homeward pennant billowed and rose from the mainmast, over the mizzen, and out to the stern and its frothy wake—like a silk hand stretched across the waters.

14 Campbell (1877) 365.

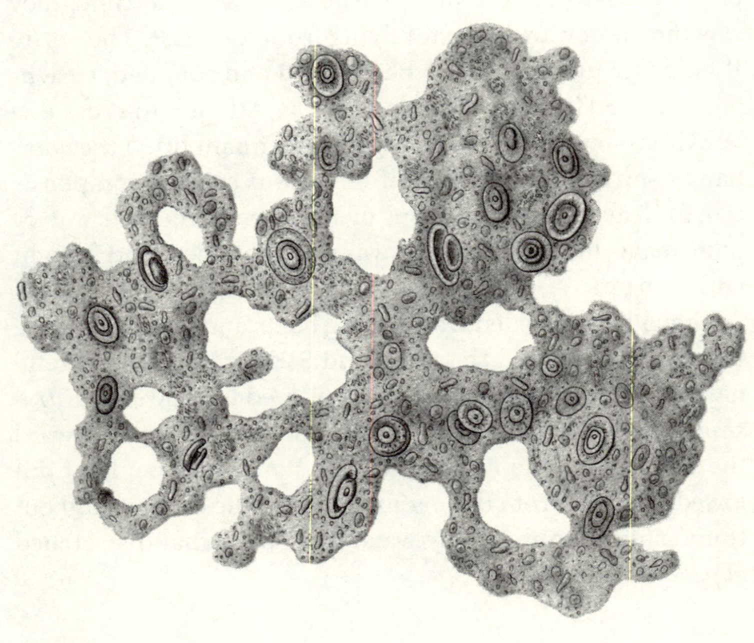

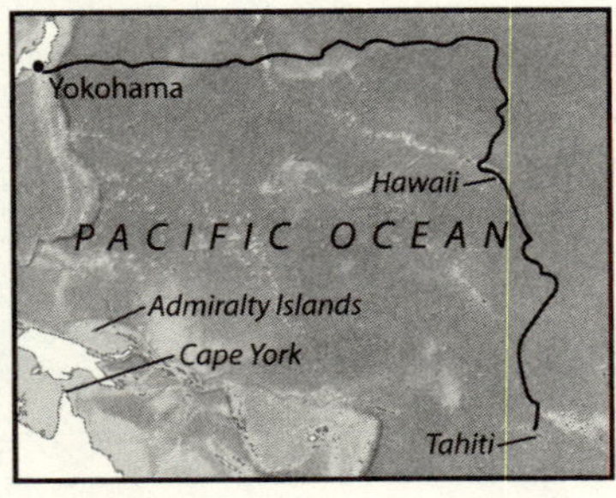

FIGURE 11.1 *Bathybius haecklii*, the phantom seafloor slime exploded as a myth by HMS *Challenger*. Ernst Haeckel, *Jenaische Zeitschrift für Medizin und Naturwissenschaft*, vol. 5, plate 17 (1870). (Inset) HMS *Challenger* Legs 10 and 11. Japan to Hawaii to Tahiti. May–September 1875. F

CHAPTER 11

Death of a Naturalist

Lō'ihi Seamount, Hawaii
18°55′ N, 155°16′ W

The Challengers had savored their last days of shore leave in Japan, except for Director Thomson, who faced professional humiliation at the hands of his own staff.

It is not too much to say that HMS *Challenger* had been sent out from England in quest of the origins of life, believed to exist in the form of a sticky film adhered to the floor of the world's oceans. At the 1868 meeting of the British Association for the Advancement of Science (BAAS) in Norwich, Thomas Huxley—Darwin's close colleague and a champion of the *Challenger* expedition—had introduced the enigmatic substance to the public as *Bathybius haecklii*, in honor of his fellow Darwinist Ernst Haeckel. Huxley pronounced Haeckel's "homogenous slimy atom" to be mobile and capable of self-generation. *Bathybius* was, Haeckel summarily claimed, the simplest form of life "which can well be imagined."[1]

The rage for *Bathybius* had begun a full decade before. Trawling the waters of the Mediterranean Sea off Nice, Haeckel's attention had been caught by "a transparent, globular particle of mucus" less than a millimeter in diameter. Without structure of any kind, the microscopic blob swam freely in the water and appeared to reproduce itself by "spontaneous fission." Haeckel declared it a new form of "Monera," neither plant nor animal, belonging to the microorganic kingdom of

1 Haeckel (1869) 34, 28.

Protista he was first to describe. The presence of pre-cellular Monera in vast populations in the sea offered, Haeckel argued, "a mechanical explanation . . . of entire organic nature." Here was the very bottom of the global food chain—and life's original expression—spontaneously generated in the watery deep.[2]

Haeckel's sensational essay on the Monera inspired Huxley to revisit deep-sea samples in his own possession—glass jars collected by HMS *Cyclops* in North Atlantic waters in the late 1850s. To his astonishment (and delight), he discerned under the microscope "innumerable lumps of a transparent, gelatinous substance" without nucleus, membrane, or any other characteristic of cellular life—just as Haeckel had described. The German had found evidence for the existence of Monera in the Mediterranean. Here was the same protoplasmic mass embedded in the mud of the Atlantic Ocean in samples from the Irish Sea to the Newfoundland coast. Huxley wrote straightaway to his collaborator, who did not hesitate in celebrating their discovery. *Bathybius haeckelii* vindicated the tantalizing speculation among Western philosophers since the Greeks that all life, human included, originated in the sea.[3]

Evidence for *Bathybius*'s antiquity and global extent accumulated rapidly. William Carpenter, reporting on ocean floor sediments from the North Sea in HMS *Lightning*, hypothesized that *Bathybius* "has existed continuously in the deep seas of all Geological Epochs." Then Wyville Thomson himself, the following year aboard HMS *Porcupine*, found in mud samples from the Bay of Biscay evidence of a "soft gelatinous organic matter" on the surface, like a sticky, translucent skin. When he added alcohol to the jar, the jelly matter separated into super-fine flakes that, under the microscope, formed "an irregular network of matter resembling the white of egg." It was

2 Haeckel (1869) 31, 223.
3 Huxley (1868) 205.

undoubtedly *Bathybius*, and Thomson publicly endorsed the new genus. "There can be no doubt," he wrote just months before *Challenger*'s departure, "that [*Bathybius*] manifests the phenomena of a very simple form of life." Darwin had confined his radical theory to the origin of new species and had not speculated on the beginnings of life itself. Now a second scientific revolution was in the offing, one that set all living being on Earth on a material, not theological, foundation.[4]

But *Bathybius*, as the putative origin of life, was destined to be short-lived. *Challenger* herself spelled its doom. The expedition had been supposed to scoop Haeckel's primordial soup (*Urschleim*) from the four corners of the world's oceans, thereby confirming it as the "material substratum of all life-phenomena." But in the course of nearly three years dredging the seafloor, during multiple crossings of the Atlantic, the Southern Ocean, and now in Pacific waters, *Bathybius* had eluded them. The primitive slime was nowhere to be found in the cold ocean mud. This failure inspired frustration, initially. Then, when Moseley, Murray, and Willemoes-Suhm—all experts at the microscope—delivered no results from sample after sample, serious doubts crept through the workroom. Thomson was baffled and feared the worst.[5]

Now, the death blow. Murray had come to him at their Yokohama hotel in a state of excitement. John Buchanan, their chemist, had definitive proof that *Bathybius* was an illusion—it did not exist. The next BAAS meeting was to be held later that month in Cambridge, where *Bathybius* was sure to be a talking point. Murray and Buchanan argued the urgent need to send their results to England—exposing Huxley's mistake—via the next mail boat from Yokohama. For *Challenger*'s scientific director, the situation was awkward in the extreme. Thomson understood well that Huxley's mistake represented, for Murray

4 Carpenter (1868) 191; Thomson (1873a) 410–11.

5 Haeckel (1869) 28.

in particular, a golden opportunity. Such a coup might launch the young naturalist's career.

Bathybius, it turned out, was easily debunked. During *Challenger*'s tedious northwards haul from the Admiralty Islands to Japan, Murray had provided Buchanan with old sample jars of mud soaked in alcohol, as per Huxley's storage instructions. To Buchanan's surprise, under the microscope these samples appeared to contain tiny flakes of the "jelly-like protoplasm" for which his colleagues had been so fretfully searching. They agreed it was not possible this evidence for *Bathybius* had been missed on initial examination, so Buchanan undertook a series of experiments to solve the mystery. He first evaporated the liquid to isolate the organic matter—without success. Suspicions aroused, he and Murray compared the older jars with more recent mud samples that had no preservative spirit added, and no *Bathybius*. After a quick analysis, the entire theoretical edifice collapsed. Huxley's substratum of life was not organic at all but rather an accident of the laboratory. *Bathybius* was nothing more than ordinary calcium sulfate precipitated by the infusion of alcohol in seawater.[6]

Presented with shocking evidence of this schoolboy error, Thomson nevertheless felt he had no option but to deny his colleagues' request to publish. According to *Challenger*'s professional hierarchy, Murray and Buchanan stood well below (future president of the Royal Society) Thomas Huxley—and himself. Priority must go to safeguarding the senior men's reputations as best he could. He drafted a quick, regretful letter to Huxley—"I have some little hesitation in writing"—to deliver the bad news. It reached London a month later, whereupon the mortified Huxley contacted the editors at *Nature* to initiate damage control. Buried in a short paragraph at the conclusion of an otherwise innocuous report, he declared *Bathybius* a myth and assumed responsibility for the "mistake, if it be one."

6 See Buchanan (1876) 605; Thomson (1873a) 408.

The *Bathybius* episode was arguably the greatest embarrassment of Huxley's long public career—and Thomson's shorter one. Years later, enemies of the man known as Darwin's bulldog were still bringing it up as proof of the folly of materialist speculations on life's origins.[7]

Riding a gusty breeze out of Yokohama, the *Challenger* philosophers were done with *Bathybius* but not with the origins—and ends—of life. An infirmary steward named James Macdonald was found poisoned the morning after their departure. The Scotsman was known to be a steady hand at sea and a filthy drunkard on land. Whether the prospect of a monthslong Pacific crossing drained his will to live or he had, in a stupor, reached for the wrong bottle on the shelf, was not inquired into.

With the Pacific Ocean again spread before them it its unchanging immensity, the ship quickly resumed a familiar, plodding routine. Their memories of Japan's colorful temples and alleyways full of charms soon took on the insubstantiality of a dream. Nothing changes, complained Jean Jacques Wild in his diary, "it is the same blue restless sea; the same sky, with its sun, moon, stars, and scudding clouds."[8]

Conditions were by no means pacific. Storms brewed about the ship and a persistent low mist made the deck and its equipment dangerously slick. They lost several trawls overboard and a full eleven miles of rope. But sounding could not be interrupted at any cost. A new telegraph cable had been earmarked for the San Francisco–Tokyo seafloor, which the *Challenger* staff were required to map. The roughness of the volcanic ocean bottom here in the western Pacific, strewn with fossil-rich pumice boulders, indicated how difficult that specific human intrusion on the deep sea would prove. But orders were

7 Quoted in Rehbock (1975) 528; Huxley (1875) 316.

8 Wild (1878) 159.

orders. The bluejackets could be seen at all hours hunched over on deck, splicing new rope brought up from the hold.

Though the homeward pennant had billowed from the mainmast at their departure, entire oceans and continents still stood between *Challenger* and home. So Wyville Thomson was surprised when his junior naturalist Suhm informed him that he planned to spend their long eastward transit to the Hawaiian Islands taking stock of his collections in preparation for their return. Almost three years out, homesickness aboard *Challenger* was epidemic.

Stacked in Suhm's cramped cabin were his German diary—two volumes of zoological notes in close, crabbed print—and its abridged English version; a great pile of drawings including newly discovered worms of the deep sea; sketches showing the bodily maturation of Antarctic krill; and a remarkable sea slug with protruding eyes on stalks. Looming over it all was the leatherbound bulk of HMS *Challenger*'s Station Book—the awesome responsibility for which had been awarded Suhm by the professor at the voyage's outset.

Across 40,000 miles of the world's oceans, through enervating heat and bitter cold, in mountainous seas and dead calms alike, the young German had kept an itemized record of the dredge's thrice-weekly contents over ever-varying beds of volcanic ash, clay, and planktonic oozes. In recent months, he had been prone to seasickness and an intermittent tropical malaise that included painful boils across his body. An unsightly red inflammation still marked his forehead. This had made daily entries in the Station Book an almost impossible strain. But the better health he had enjoyed since Japan, and the publication potential of the untold pages of scribble on his desk, had rekindled Suhm's ambitions. The older scientists thought his indifference to sleep commendable.

Seated at his desk crowded with papers, the enormity of his workload struck Suhm with new force. He had half a dozen statistical tables to provide the Royal Society; research letters to send to his mentor, Professor von Siebold, in Munich; and his English reports, neglected during a period of illness in the Philippines, to bring up to date. Composition in English—technical, scholar's prose for professional eyes—was never as easy as his native German. And so he balanced English clauses late into the night, having already spent long hours in the workroom bent over his microscope committing the contents of the day's dredge to the posterity of the Station Book. Then all at once, a week out of Yokohama, an extraordinary chance upended his well-laid plans—an opportunity to engage the attention of Charles Darwin himself. There could be no hesitation. Suhm shoved everything else aside and threw himself into the study of an unusual open-ocean barnacle.

The thirty-fifth parallel from Japan to the Hawaiian Islands lies between tropical and temperate zones. During the early days of their transit, the contents of *Challenger*'s tow net had a corresponding mixed, indifferent character. The radiolaria, copepods and other surface plankton were familiar to Suhm from colder waters, while the creatures he was most interested in for publication purposes—the larvae of tropical shrimp species—had inconveniently vanished. Then, in the course of a single summer's day, June 21, the temperature of the sea dropped 5°F (2.8°C), turning the surface waters into a graveyard. The ship's prow cleaved through thick carpets of dead plankton in alternating red and white patches according to species—all victims of the sudden inflowing cold current from the north.

Amid the carnage, an entirely unexpected survivor swam into their ken. Floating balls of barnacle larvae bobbed along the ship's side, complete with tails and thorny spines, and adhered to the surface of thousands of dead *Velella*. These jelly-like hydroids were well-known to the crew as "by-the-wind sailors" for their resemblance to tiny yachts, with a stiff, gelatine sail by which they navigated the open ocean (entirely like their giant counterpart HMS *Challenger*). But there was still more to this remarkable parasitic compound. Encrusted on the surface of the barnacle larva balls were tiny shell-less gastropods, a flesh-eating louse, and a sea mouse (really a worm) named inexplicably for the goddess Aphrodite.

Suhm quickly established that these balled-up larvae belonged to a species of oceangoing "buoy-barnacles" described two decades prior by Darwin—*Dosima fascicularis*. For days, *Challenger* sailed through entire streams of full-fledged *D. fascicularis*, plucked with ease in the tow net. Back in the workroom, he encouraged these mature captives to reproduce. Soon an entire barnacle life cycle—egg, larva, chrysalis, and adult—lay before him in original splendor. Even Darwin himself—world authority on the Cirripedia—had never had the benefit of live barnacle dissections at sea to examine their stunning metamorphoses in real time. Suhm wrote to Professor Siebold with barely suppressed excitement. Providence had been kind enough to bring him to this remote corner of the great Pacific and a chance at renown (it would be the last such kindness).[9]

The *Challenger* philosophers had scraped a cluster of gregarious goose barnacles from the shell of a giant crab in Japan's Inland Sea but were unable to identify them. Two years earlier, Wyville Thomson had better success with his discovery of a new species of deep-sea barnacle east of Bermuda. Half a dozen *Regioscalpellum regium* had spilled on the ship's deck in various degrees of attachment to a slippery ball of manganese, the origin of which utterly baffled them (and would result in one of *Challenger*'s most important publications). This female *Regioscalpellum* barnacle was notable for its large size—two and a half inches long—and for the little party of parasitic males attached to it.

9 Willemoes-Suhm and Darwin knew the barnacle as *Lepas fascicularis*. A century later, it was reassigned to the nearly related genus *Dosima*, of the common family Lepadidae.

Darwin had responded gratefully to Thomson's description of the new *Scalpellum* in the pages of *Nature.* He saw it as an opportunity for reflections on animal sexuality and evolution, a full quarter century after his own barnacle research had revolutionized the field of invertebrate biology. Darwin had never intended to devote eight years of his career (1846–1854) to the Cirripedia. But the ubiquitous barnacle—with its near-infinite variations and bizarre sexualities—proved the perfect subject for a closet evolutionist. The *Challenger* hands had spent backbreaking days at the dry dock in Yokohama scraping the familiar acorn barnacles, adhered in their thousands to the ship's hull. These creatures Darwin had itemized minutely from samples sent to him from around the globe. Suhm's *Dosima*, however, belonged to the more enterprising stalked or "goose" barnacles, which colonize floating objects of the open ocean—hitching a ride on anything from kelp, snails, and dead *Velella*, to logs, sharks, and whales.

Darwin was especially fascinated by the cirripeds' ingenuity in reproduction. Most barnacles were hermaphrodites capable of self-fertilization, but he discovered in others a kind of parasitic dimorphism—with "dwarf" males (little more than squidgy sacs of sperm) embedded in the female (Thomson's new *Scalpellum* was one such). Some barnacles appeared to change sex presentation during the course of their life cycle, while hermaphroditic "males" boasted a penis up to eight times their body length, designed to douse the tiny, plate-enclosed cavities of adjacent "females."[10]

This generative flexibility, confirmed by the fossil record, suggested to Darwin a sequence of incremental change and adaptations *over time*—evolution, in other words. Hermaphroditism in barnacles had given way to dimorphism, dwarf

10 As recently as 2013, it was discovered that some barnacles are also capable of so-called spermcasting, which implies fertilizing females at remote distances by spraying at random into the surrounding water. See Barazandeh et al. (2013).

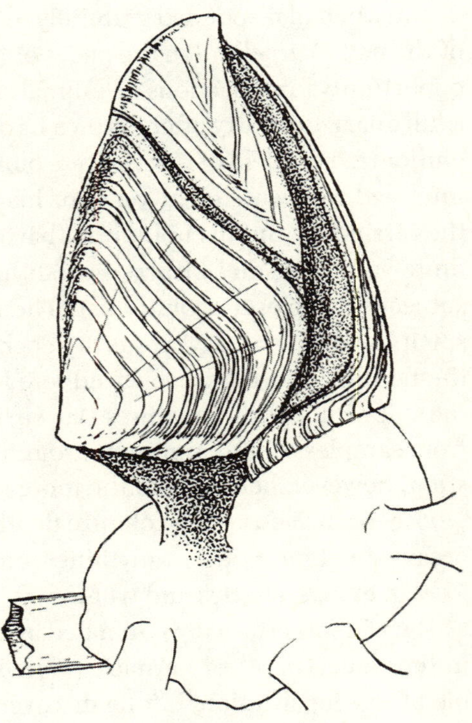

FIGURE 11.2 Open-sea stalked barnacle *Dosima fascicularis*. From *U.S. National Museum Bulletin* 60, plate 9 (Smithsonian Institution, 1877).

males, and other baroque spawning strategies, as ocean conditions changed and new competitors emerged. Reproductive success determined the natural order—and the origin of new species. Indeed, Darwin's baseline principle of "fitness" is nowhere better exemplified than in *D. fascicularis*, which brave brutal open ocean conditions en masse in their irrepressible search for new niches. Even in the 1870s, barnacle ecosystem dominance was proverbial. Specimens of *D. fascicularis* had already been found the watery world over.

Following Darwin's example in his classic barnacle monographs, Suhm focused his attention on the weird metamorphoses of his goose barnacle. It was his somber duty, first of all, to correct his senior colleague, biologist Anton Dohrn, who had mistakenly awarded this larva its own genus, which he called *Archizoëa*. But barnacles have a long history of deceptiveness. The original fellows of the Royal Society clung to the folk tradition that barnacles were goose eggs. Carl Linnaeus, giant of the Enlightenment, had reclassified them as mollusks, while Lamarck considered them adjacent to worms. Only in 1830, when Irishman Vaughan Thompson observed the translation of free-swimming larva into plated rock dwellers, did the familial resemblance to crustaceans become apparent.

Under the microscope, Suhm observed the transparent, oval-shaped eggs of *D. fascicularis* multiply their cells to form distinct appendages—antennae, a tail—enveloped in a thickening, granular skin. Emerging from the egg, the larvae sported a rudimentary spine nested in segments "like the tubes of a telescope." Horns, glands, eyes, mouth, and anus followed, all crammed inside a protective carapace.[11]

For Darwin, *D. fascicularis*'s signature adaptive wonder was its dual-use cement, secreted via its glands and head—designed both to attach itself to its host *and* provide all-important buoyancy. Suhm could confirm that the *Dosima* pupae, clinging in numbers to the jelly-like *Velella*, did not sink their fragile raft but acted rather as serendipitous floats. Finally, with its upside-down head firmly glued, its plated shell in place, and a row of feathery, exposed feet to scoop in food, the *Dosima* barnacle's picaresque path to maturation was complete. Suhm finished his paper with unfamiliar feelings of professional triumph. He mailed it from Honolulu to the Royal Society in London, where it arrived two weeks after his burial at sea.

A flock of young albatrosses had accompanied *Challenger* east. Their plumage was still dark with only the suggestion of mature whiteness about the neck and head. The frustrated anglers aboard baited them with hooks. They spared one from

11 Willemoes-Suhm (1876a) 130.

the stew pot in order to paint it white, hoping to track it. The great birds fed complacently about the ship in social numbers until, one day toward the end of July as *Challenger* made her southing toward Hawaii, a tropical shark appeared and the cold-loving albatrosses vanished. It was as if the respective rulers of the ocean air and waters had divided the kingdoms between them along an agreed-upon temperature boundary.

They crossed the shimmering, unseen line into tropical heat, and promptly lost their friendly breeze. The Challengers dawdled into Honolulu Harbor ten days later parched and exhausted. After weeks at sea, the land seemed like an accident, a curiosity of creation on an otherwise aqueous globe. The rising volcanic islands appeared barren from the deck, starkly different from the lush tropical coasts they had been accustomed to. The Americans had brought prickly pear with them to Hawaii's islands, where it had spread like the plague across the bare rock and thin soil. Thickets of guava bush—another invader—surrounded the town, whose streets and buildings had a fully American appearance: regular, white-painted houses with wrap-around verandas to encourage the circulation of air. The indigenous people of the islands had died in epidemic numbers in recent years. Their extinction was inevitable, the Challengers were told, which Henry Moseley interpreted as license to desecrate local burial grounds for skulls and other human remains.[12]

For three years, the Challengers had with great effort dredged up cubits of volcanic red clay from the ocean's bottom. Glassy shards and pumice grains from a thousand eruptions dating back millions of years and covering most of the submerged Earth sat quietly in jars on the workroom shelf or crated up in the hold. Now at Hawaii, they did not pass up

12 Undoubtedly the least creditable of the *Challenger* expedition's several hundred publications is "Report on the Human Skeletons Collected," authored by Sir William Turner. *Scientific Results: Zoology* (1884) vol. 10.

the opportunity to examine a living source of the red clay: Mauna Loa, the largest active volcano on Earth not hidden beneath the ocean.

The approach to the famous crater barely registered as an ascent, but it was endless. At last, the lines of bush on either side thinned then vanished. "On our left lay a vast depression bounded by blackness," Lord Campbell remembered. A white vapor from the earth enveloped them, drifting toward the crater. At the edge, they saw the moon in the still, black sky brilliantly reflected on the vast lava bed. The silver surface of the volcanic lake was cut through with electric pink lines leading to the crater wall. There the molten lava surged against the cliffs in fountains of fiery spray. Even the sun itself seemed overshadowed, sinking mildly behind the rim so that in the falling darkness the white-hot lava glow grew more intense, illuminating the cliff walls with crazy shapes. They stood amazed by the furious display. The "tumbling, heaving, crimson molten lava" defied their senses and the logic of the ordinary world. The great lake burned without flames; its crashing waves, enveloped in hissing gases, were like an ocean dipped into the fires of hell.[13]

Suhm was in no condition for a volcano ascent, however tremendous. On *Challenger*'s arrival in Honolulu, he made his priority to recuperate for the long passage home. He refused social invitations, pleading the excessive heat, and spent hours on the balcony of his guesthouse surrounded by its lush garden filled with introduced plants. He ate pineapples, bananas, and mangoes, and imagined his vital energies returning.

He ventured out one day to lunch with a Hawaiian prince, who grilled him with questions about the high cost of living in Europe. Together they visited the mausoleum built for his cousin, a prince of high rank buried in a coat made from the yellow feathers of the rare Hawaiian mamo bird (*Drepanis*

13 Campbell (1877) 391.

pacifica), a native breed of honeycreeper. The brilliant coat was the work of generations and the pride of the Hawaiian court. Inspired, Suhm took to exploring the wild fern forests behind the town in quest of a mamo for his own collection. He saw extravagant birds and plants he had seen nowhere else in the world. He cut his way through the "gray-green shimmering arbor" until he arrived at the banks of red flowers he had been assured would attract the *D. pacifica*. And there, exhausted by the terrible forest heat, he captured his prize—a bird with yellow feathers fit for a dead prince (the mamo is now extinct).[14]

Ernst Haeckel had theorized that the embryonic development of an individual animal recapitulated its evolution as a genus. Suhm's yellow-feathered mamo, for example, had evolved from a simple common ancestor as one of a host of complex, variant forms, both bird and nonbird. And these secrets of its descent were contained within its brittle, speckled egg. According to Haeckel's theory, the barnacle larvae swimming under Suhm's microscope in the *Challenger* workroom likewise doubled as ghost images from deep time past when Earth's primitive ocean was populated by incipient, single-celled life.

For a giddy year or two, Haeckel had convinced himself that *Bathybius* was the most primitive form of all being—the origin of life itself. That was before Murray and Buchanan joined forces to debunk him. But truth was, he had already made Victorian science's best guess at that eternal question by devising the first tree of life to include microorganisms. Bacteria—first identified and named in 1872 as a genus of Haeckel's protistan Monera—are the most likely candidates to have provided an original fusion for cellular life.

The philosophers of HMS *Challenger* were well acquainted with the ubiquitous blue-green algae now classified as cyanobacteria. But in all their voluminous writings, they rarely

14 Willemoes-Suhm (1877) 177.

employed the unfamiliar word *bacteria*. An entire universe of microorganisms was too tiny for their microscopes and was thus hidden from them. It is, in one sense, tragic. The Victorian scientists sought answers to deep questions in evidence they were not equipped to find, and without a language to interpret it if they did. A study carried out in the late 1990s off the coast of Hawaii revealed, for the first time, the colossal presence of bacteria in the oceans—fully 90 percent of all cellular biomass in the upper 150 meters of the water column. En route to Hawaii, the Challengers were not aware of the literal oceans of bacteria they sailed along, the true substratum of life that would soon claim one of their number. The workroom microscopes revealed a teeming microbial galaxy of the sea. But their magnification was too primitive to penetrate to the bacterial legions, both bounty and threat.[15]

Based on current understanding, what were the oceanic conditions on early Earth that enabled the ubiquitous bacteria to first appear? One theory spotlights hydrothermal vents.

In the 1970s, during research missions to mark the centenary of the *Challenger* expedition, scientists stumbled upon heat-producing fissures in the ocean floor off the South American coast. The discovery revolutionized modern marine biology—the vents support vast communities of unique organisms—while also bringing the ocean floor to the attention of the mining industry and venture capitalists.

Hydrothermal vents send plumes of superheated water from magmatic chambers beneath the seafloor. As the plume cools, dissolved metals in the water column bind into sulfide particles and sink to the bottom, eventually forming tall,

15 Breakthroughs in marine microbiology were not far in the future, however. Motivated by the success of HMS *Challenger*, the government of France launched a series of oceanographic expeditions in the early 1880s. A colleague of Louis Pasteur's, Adrien Certes, first identified ocean bacteria in sediment and water samples collected by the *Talisman* and *Travailleur*. See Adler and Dücker (2018).

chimney-shaped structures—called "black smokers"—replete with copper, zinc, gold, and silver, all at far higher concentrations than terrestrial deposits. Black smokers are living relics of our primordial planet, dating from the first appearance of liquid water on Earth more than four billion years ago. They also contain elements vital to core products of the global high-tech economy such as smartphones, LEDs, and batteries for electric vehicles.

The "energy releasing chemistry" that, over the course of millions of years, produced sulfide deposits at hydrothermal vents might plausibly have given rise to the very first biochemical pathways of life on Earth—in bacteria. Researchers from the University of Hawaii, examining microbial mats from a hydrothermal vent at the Lōʻihi Seamount, have identified clusters of Proteobacteria with ancient genes capable of subsisting upon sulphur compounds in a lightless, superheated environment—a clue to life's genesis. The Victorians' much-trumpeted *Bathybius* turned out to be an accident of chemistry. Chemical "accidents" at hydrothermal vents three billion years ago plausibly altered the destiny of an entire planet.[16]

At different stations across the Pacific, *Challenger* collected bottom sediments with unusually high levels of manganese and iron. It was a hundred years before these enigmatic samples were correctly interpreted as indicators of hydrothermal vents (Murray proposed undersea volcanoes—a good guess). No one onboard knew it, but *Challenger*'s first station south of the Hawaiian shore stood only a few miles to the northeast of the submarine Lōʻihi Seamount, within the same circuit of beach-bound waves. In 3,000 fathoms of clear blue water, they dredged up sands almost entirely composed of disintegrated volcanic rock. Glassy threads signified its origin in craterous fires—or the nearby hydrothermal vent. The further they sailed

16 Martin et al. (2008) 812.

from the islands, however, the vitreous minerals diminished in size and volume until the ambiguous bottom deposits of clay, radiolaria, and *Globigerina* shells fully resumed. They were once more on the open ocean and divorced from the terrestrial order of things.

By *Challenger*'s last summer, 1875, dozens of her original crew had deserted at ports across the globe. The death toll, meanwhile, counting the unlucky James Macdonald, stood at eight. At the time the radiolarian ooze reappeared in the dredge, south of Hawaii, everyone on board was aware that another life onboard hung in the balance: the foreign scientific gentleman universally known as Suhm.

Oxygen-producing plankton provided for every second human breath taken aboard *Challenger* but also harbored threats for terrestrial interlopers on the seas. Somewhere in the constant traffic of buckets of water spilling across *Challenger*'s deck, in the daily hand-sorting and sealing of specimen jars cloudy with marine matter, and sitting for hours in work clothes drenched with scum, Suhm's raw, unseamanlike skin, still not recovered from its recent tropical eruptions, met the flesh-hungry bacterium *Streptococcus*.

John Buchanan, Suhm's closest friend onboard, found it inexplicable that he should be attacked by disease after a full month at sea. The weather, throughout his short illness, could not have been better for the care of an invalid. A mild sun shone, while the light breeze from the east was just sufficient to fill the sails. Erysipelas first announced itself in a prolonged shivering fit in the first week of September. Forced reluctantly from his desk, Suhm reported to Dr. Crosbie, who ordered a sick tent be erected on deck to take advantage of the cool, fresh air.

Until his final hours, the young morphologist maintained his strength and appetite. But once the inflammation spread across his head and neck, he sank quickly. The rapacious bacterium, introduced to his bloodstream via a burst boil on his face, had run amok. Suhm fell into a delirium and began rambling in German. For a full day while they floated across the flat, landless disc of the Pacific Ocean, he imagined himself hiking the Bavarian Alps. In a letter to his grieving mother, Buchanan assured her that her son had borne "his terrible sufferings with the greatest courage."[17]

For the *Challenger* naturalists, who felt assured of his total recovery until nearly the last, the death of their colleague was a shocking blow. He was the youngest of them; his death seemed a reversal of the order of nature. Director Thomson himself helped bear the body to the ship's rail for deposit into the sunlit Pacific. According to Navy tradition, *Challenger* resumed normal services once the captain closed his prayer book. The deckhands dismantling Suhm's sickbed indulged their usual chatter and pranks. The grieving scientific staff were not accustomed to this naval show of indifference to death and secretly resented the men for it.

The last publication of Suhm's short career was his life-cycle description of the goose barnacle *D. fascicularis*, which *Challenger* had observed in passing at multiple ports and stations, but which absorbed their full attention in waters west of the Hawaiian Islands for a week in the summer of 1875. For Suhm, the wandering, protean barnacle offered a classic instance of Darwinian adaptation and success.

Today's Anthropocene seas are a scene of accelerated Darwinian change, with a novel cast of winners and losers. Microbes of the kind that killed Rudolf Willemoes-Suhm are rapidly filling niches in depleted habitats, spreading disease among fish and crustaceans, and bubbling up in the toxic algal blooms that clog the world's urbanized coastlines.

Barnacles, too, are profiting from the worldwide breakdown of ocean ecosystems. As a sailor "by the wind," the storied *D. fascicularis* is naturally prone to running aground. A 2011

17 Willemoes-Suhm (1877) 179.

study of debris washed up on beaches along the South African coast found a legion of Suhm's goose barnacles adhered to bits of trash. Barnacles have successfully populated the global marine environment since Cambrian times, managing climate change and mass extinctions along the way. Now, a worldwide scum of floating, human-made plastic is "drastically increasing" *Dosima*'s opportunities for oceanic expansion. It's a barnacle's world, and will continue to be.[18]

18 Whitehead et al. (2011) 635.

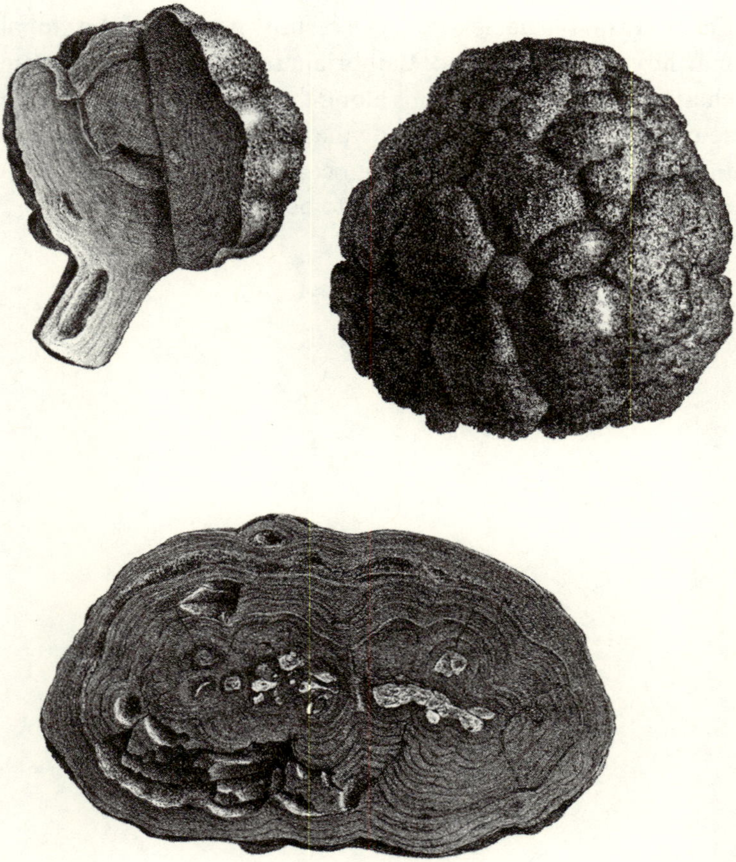

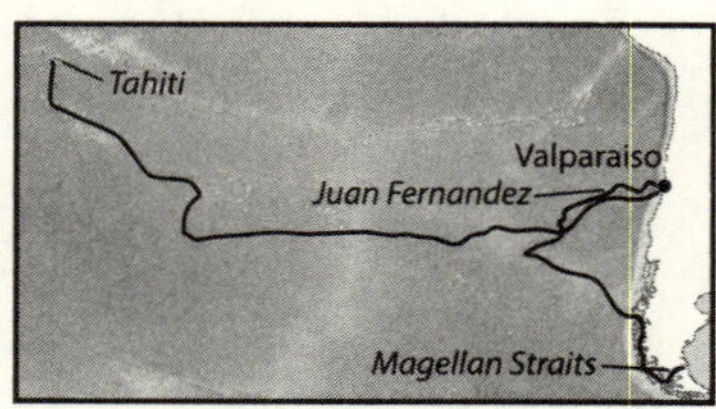

FIGURE 12.1 Varieties of manganese nodules collected by *Challenger* and analyzed for the first time by John Murray in his landmark volume *Deep Sea Deposits* (1891). (Inset) HMS *Challenger* Leg 12. Tahiti to the Magellan Straits. October 1875–January 1876.

CHAPTER 12

Message from the Cosmos

Valparaiso, Chile
33° S, 71°6′ W

Challenger's decks ran slick with warm rain—the ship marooned in its own waste. Weeks adrift in the tropics proved to be a great social equalizer. Scientists, officers, and men alike suffered in the moist heat. Then a light trade wind out of the southeast filled *Challenger*'s sails and carried them across the Equator for the fifth time in three years. Equatorial currents—a teeming highway of life cutting through the broad Pacific wastes—delivered buckets of plankton and lit the ocean surface nightly with a phosphorescent glow. Station 271, the morning after crossing the line, proved the "richest source" of surface radiolaria of the entire voyage.[1]

The top of the food chain was equally well-represented by cruising sharks and whales, and by rubbish left by seafaring humans. They sliced open one shark's stomach to find bits of pork and old rope. Then, on September 16—two days after the solemn burial of Rudolf von Willemoes-Suhm—the dredge coughed up a harpoon spear among a heap of whale teeth. A mighty sperm whale had met its end at that very spot in the Pacific vastness, though it escaped being carved up and drained for its oil. In the relative nutritional desert of the open Pacific, a hungry deep-sea population—from sharks to mussels to gutless worms—relies on a steady supply of mammalian carcasses for its existence. Given the ravages of the whaling

1 Haeckel (1887b) ii.

industry in the Victorian period, those communities must have already faced near-total collapse. *Challenger* arrived too late on the scene to record their original extent and character.

While his colleagues studied the enormous vertical acreage of the water column and its living inhabitants, John Murray focused his attention on its horizontal extremities: the sea surface and floor. Boating about under the harsh equatorial sun with his trusty net, he gathered jars full of living plankton that matched the cold, *Globigerina* ooze of dead shells scooped from the bottom by the dredge. But as the waters deepened and acidified, the tiny calcareous creatures disappeared from the bottom deposits. Jewel-like radiolarians, made of robust silica, dominated the mud. Then the contents of the dredge changed character again. Further south, beyond the nutrient-rich currents, all planktonic ooze vanished, leaving only the enigmatic red clay. For long hours, Murray pondered what its purpose might be—beyond chilling a bottle of champagne. Now it delivered a further puzzle.

In the first months of *Challenger*'s journey, off the Canary Islands, Murray had made a perplexing geological discovery. Embedded in the red clay coating the seafloor, he found tiny balls of black manganese alongside sharks' teeth coated in the same black stuff. Now, in the South Pacific approaching Tahiti, the dredge began disgorging bucketsful of manganese nodules the size of cricket balls. Cutting one open, he found a shark's tooth embedded in its heart within concentric circles of volcanic glass, light-colored in the center shading to dark at the rim. These manganese layers, polished to a brilliant shine, represented an astonishing visual record of deep time.

Murray's manganese nodules—now called polymetallic nodules—are eons in the making and require a stable aquatic environment to form. Dissolved metals in seawater precipitate around a bone, tooth, or shell at the stately rate of ten millimeters per million years. As long-term residents of the deep sea, manganese nodules provide a unique habitat for seabed life. Crusted algae, polyps, sea squirts, and worms stick to the mottled nodule's surface. Other community members are less clingy. Recent video from submersibles has shown an ever-shifting parade of deep-sea fish, prawns, sea cucumbers, and brittle stars in and among the dotted carpet of manganese nodules on the sandy floor. The nodules are more than scenic background to this rich benthic biosphere. New research points to the manganese balls as a vital source of "dark oxygen" in the deep sea, a product of seawater electrolysis.[2]

At multiple stations across the central and southern Pacific Ocean, *Challenger* recorded polymetallic nodules "in extraordinary abundance." South of Hawaii, the nodule haul gradually increased in volume culminating in a bumper crop of half a ton, ranging in size from a marble to a head of lettuce—enough to fill two casks. Murray packed samples of the enigmatic balls in a crate, with a pile leftover as souvenirs for the officers. His geologist's instinct told him that money might one day be made from this deep-sea harvest. A century and a half later, that day has arrived.[3]

The fantasy of ocean floor industrialization dates to the *Challenger* era: in the words of Captain Nemo—"I'm telling you that reserves of zinc, iron, and silver exist at the bottom of the sea, and their exploitation would be eminently practicable." Now, in the twenty-first century, with terrestrial reserves of industrial minerals—nickel, copper, cobalt, and rare earth metals—in steep decline, serious attention has turned to the ocean floor as a last chance mining frontier.

The Penrhyn Basin first plumbed by John Murray aboard HMS *Challenger* covers an area of 2 million square kilometers

2 Sweetman et al. (2024).

3 Thomson and Murray (1885) 774. In addition to manganese nodules, deep-sea mining interests are focused on extraction of polymetallic sulfides at hydrothermal vents and cobalt-rich manganese crusts found on seamounts, both first sampled by HMS *Challenger*.

(770,000 square miles) spanning from the Cook Islands to Tahiti and north. The basin floor is littered with billions of polymetallic nodules containing high concentrations of cobalt and rare earth metals vital to the production of high-tech products, such as smart phones and electric cars.

Unsurprisingly, the Penrhyn Basin and other Pacific deep-sea sites—most notably the Clarion-Clipperton Zone, an area the size of Europe—have attracted a rush of mining license applications to the International Seabed Authority (ISA), the UN agency charged with management of the ocean floor outside national waters. In the past decade, the ISA has issued dozens of exploratory permits to companies backed by China, Russia, and other national governments. Amid the stampede to industrialize the seabed, a moral question of literally planetary dimensions remains unanswered: Is it defensible to sacrifice vast swaths of the deep sea, with its legion of unique plants and animals, to supply our green energy revolution?

The environmental costs of mining the deep sea almost baffle description. Harvesting polymetallic nodules in commercial quantities will require deployment of massive dredging equipment on the seafloor, operating twenty-four hours a day for years on end. The millions of tons of soft sediment churned up by mining will not only destroy the seabed habitat itself but, through the release of toxic plumes into the surrounding water column, will decimate marine animal communities for thousands of miles across the Pacific Ocean. Extreme noise and light pollution from seabed machines and their support vessels will compound the horror. In 2016, a research team revisiting the site of an experimental deep-sea mine from the late 1980s found a wasted seafloor devoid of life. For an ecosystem built over millions of years, there can be no meaningful time frame of recovery.[4]

4 See Vanresusel et al. (2016). A stark analogy for deep-sea mining impacts is the destruction wrought daily by the international fishing fleet. Commercial

Perhaps trillions of manganese nodules are spread across the world's ocean floors but are only accessible in remote places where the sedimentary outflow of continents and the perpetual snow of dying surface plankton have not buried them. A manganese nodule sitting on the ocean floor, with the tooth of an extinct megalodon at its core, represents a kind of crystal ball of Earth history.

The manganese nodule is also an emissary of the stars. Under a microscope aboard *Challenger*, John Murray noted the enigmatic presence of dozens of tiny metallic pellets in red clay sediments dredged from the ocean floor. On return to Edinburgh, he took a clutch of nodules from *Challenger*'s South Pacific leg, broke off sections, and pounded them with a mortar into a fine dust. He then took a magnet and applied it to the disintegrated matter. Attached to the magnet, he found the same tiny iron spheres. Their composition, he decided, must be literally out of this world.

"Cosmic spherules"—first described by Murray in his trailblazing *Challenger* report on deep-sea deposits (1891)—are metallic remnants preserved in the seabed that originate from the vast asteroid belt encircling the sun between Jupiter and Mars. Continual collisions in that chaotic rocky firmament release particles to remote corners of space, including our planet. Cosmic spherules are abundant but not easily recovered on terrestrial Earth. The airborne cosmic dust visible from the deck of HMS *Challenger* as a cone-shaped zodiacal light on the horizon after sunset was the same starstuff coating the ocean floor beneath her hull, to be scraped up by the dredge. Each of the tiny meteor fragments under Murray's microscope had melted on entering Earth's atmosphere, then cooled and hardened into a sphere. The inescapable conclusion? Hidden within the ancient manganese nodule was this

---

beam trawls have desertified coastal seafloors worldwide for the past century, which mining will replicate in the open sea.

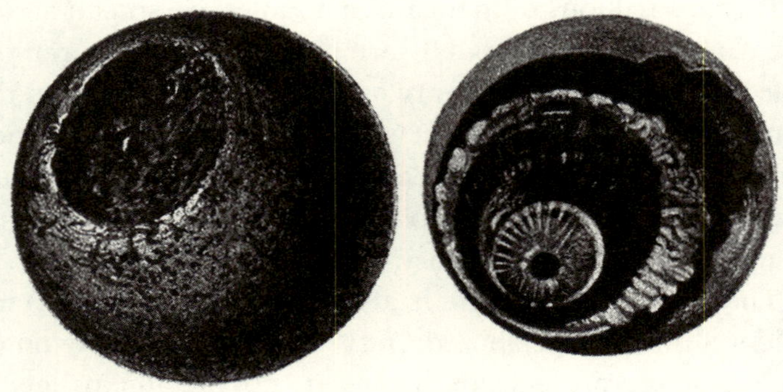

FIGURE 12.2 Cosmic spherules, visible only under a microscope. *Scientific Results: Deep Sea Deposits*, plate 23 (1891).

even more exotic time capsule, a messenger not from our planet's history but from the origins of the solar system itself, even before the Earth or the existence of the oceans that *Challenger* sailed across.

History, on an entirely different timescale, awaited them on Tahiti. From a distance, the palm-lined beaches and green mountainsides, framed by a turquoise sky, appeared like the paradise of European fantasy. But once onshore, that aura of untouched beauty faded. The scrubby guava tree, brought decades prior by an American speculator, had devastated the native forests. Colonies of rats ravaged the plantain trees, also introduced. Invader pigs ran wild in the mountains. Cattle and sheep, imported from Hawaii, were "the most miserable specimens" Henry Moseley had ever seen, while the Tahitians themselves were increasingly crowded out by Chinese and European traders of all stripes. Local French officials—conspicuously unimpressed by HMS *Challenger*—supervised plantations of coconut oil, cotton, and vanilla beans, in addition to the installment of the Christian religion and its dress codes.[5]

5 Moseley (1892) 447.

The islands of Juan Fernandez—a six-week tedious sail to the east—were likewise disillusioning. An English sailor named Selkirk had been marooned there, from whose memoirs Daniel Defoe had embroidered the famous tale of Robinson Crusoe. But the novelist had transplanted the setting of Selkirk's ordeal to the lush Caribbean, a stark contrast to the reality of Juan Fernandez—a gloomy, barren rock rising out of the empty Pacific. The main island, overrun by goats and rats, was owned by a Chilean merchant whose sole interest, they were told, was the profitable extermination of its seal population. Seal skins brought sixteen dollars apiece in the current market.

Anchored offshore of Juan Fernandez, it was possible to maintain a sensation of romance. They arrived on a stormy evening in November to the sight of a dark sea crashing the cliffs and clouds of birds sweeping across the bay. The mountains, cut with ravines, seemed "torn and broken into every conceivable fantastic shape."[6]

The sunny morning that followed brought an unexpected bounty: the best day's fishing of the entire voyage. One party took their rods and lines to the headland rocks, where they battled local sharks for the shoals of cod and bream swimming forty fathoms below. Those left onboard fared even better. Lobster pots thrown over the side delivered crayfish and conger eels by the dozen. A simple line cast across the water's surface drew up tasty crevalle jack one after the other—more than could be eaten by the entire mess. By sundown, even the captain's cabin smelled like a fishmonger's.

The blue purity of the South Pacific—their home for the last several months—had yielded little of edible interest. Now, with a great continent in the offing, the seas about them had repopulated dramatically. East of Juan Fernandez, the rich, cold waters of the Antarctic rose up from the south to meet

6 Spry (1878) 338.

them, turning the ocean surface from lifeless blue to a teeming green. With the serrated coast of Chile emerging out of the haze, they entered the most profitable fishery in the world. Gulls in great flocks flew out to greet them. "The water," wrote Murray in his diary, "is full of small living things." Also giant things. At the mouth of Valparaiso harbor, they watched as a whale and her calf breached again and again with a towering splash. Their massive bodies, twisting in the air, seemed to claim dominion of the watery element beneath—at once innocent of gravity and separated from time.

A subtle shift in undercurrents had preceded their arrival in the great Chilean fishing grounds. At the fortieth parallel, south of Tahiti, a persistent west wind forced Captain Thomson to alter course. Sailing due east toward the South American coast, the daily ocean temperature readings grew colder at a shallower depth along the same line of latitude. Even on the final day of their transit, as *Challenger* steamed north toward Valparaiso, sea temperatures at depth continued to plunge. They had joined the Humboldt Current, a submarine torrent of polar water barrelling northward along Chile's coast. The Humboldt's power was expressed beyond the southern horizon in the spectacular glaciers of Patagonia—whose climatic conditions it helped create—and in the marine life bonanza in its wake.

Indigenous beachcombers, encouraged by warming temperatures, had settled this bountiful stretch of the American coast ten thousand years ago, where they dived for mussels, speared fish in the shallows, and ventured seaward on boats made of seal skins. The subtleties of the polar current and its cold-loving fish were well-known to them. Millennia later, when new colonists arrived, Alexander von Humboldt was the first European to trace the hemispheric course of the cold current from the Antarctic circle to the continental shoulder of northern Peru, where it pivots abruptly westward to infiltrate tropical waters. This was the Southern Hemisphere's monumental answer to the Gulf Stream. Humboldt published his "discovery" in an 1846 book whose title, *Cosmos: Sketch of a Physical Description of the Universe*, captures well the Faustian tendencies of the scientific culture he represented. The current duly bears his name.

In the same passage from *Cosmos*, Humboldt describes the dialectical romance of the seafaring naturalist in the Americas. It reads as a kind of tourist manifesto for the naturalists of HMS *Challenger*. First, says Humboldt, the Pacific explorer will inevitably fix his eye "on the distant sea's rim where air and water meet." The boundless horizon, ever-slipping from sight, seems to insist on its symbolic meaning as "a sublime image of the infinite" and his own inquiring mind. This, for an alpine mountaineer, would be satisfaction enough. But sea voyaging, Humboldt insists, offers an experience of nature's wonders at multiple scales. With a single turn of the head to the busy waters around him, the ocean traveller is drawn out of self-absorption and into the shimmering aquatic world: "all the richness and variety of animal life . . . of highly organized and beautiful forms."[7]

Humboldt's formula applied neatly to the *Challenger* philosophers' experience during their long, desultory leg from Tahiti to Valparaiso. For weeks, becalmed on the Pacific Ocean in all its pointless perfection, and with the dredge empty of novelties, there seemed little to do but stare at the horizon. Then, one morning, they found themselves in the Humboldt Current brimming with life, and they were suddenly like children at an open-air aquarium.

Valparaiso, not coincidentally, is situated where the upwelling of Chilean coastal waters, driven by strong southerly winds, is most intense and productive. Lucky *Challenger* had arrived in early summer, the season of peak upwelling.

7 Humboldt (1849) 303–4 [translation modified].

Cruising about the ship with a tow net, Murray and Moseley plucked jellyfish galore from the choppy surface waters off Valparaiso. Also many novel sea cucumbers: one translucent, a body inside a body; another with the shape and color of an eggplant. They scooped up a starfish with thirty arms, a gelatinous worm with fern-like paddles, a porcelain-shelled crab, and a purple one. In a few short hours, an entire museum gallery new to European science wriggled in buckets between their freezing knees: keyhole limpets, urchins the size of a baseball mitt, and a periwinkle with attractive black-and-white stripes. More shrimps followed, also arrow worms, and a luminescent ostracod glowing with rage (*Cypridina*).

To add to their spectacular South American haul, the tow net trailing from the ship's stern trapped several mysterious tunicates—from the Larvacea group—for which evolution had designated a singular niche. Tadpole-shaped as adults, the Larvacea inhabit an oversized balloon of mucus of their own making. The rapid motion of their tails pumps ambient plankton into the tunicate's "house" to be devoured. Ubiquitous in colder waters, the Larvacea, some with balloon houses a meter wide, play a vital role in transporting surface carbon to the deep sea.

Also from the tow net emerged a slick *Cirrothauma* octopus, with a bat-like head and parachute arms. It recalled their time in adjacent Antarctic waters, as did the long-legged sea spider *Nymphon* and their jars full of tiny copepods, staple food for the entire bay community. Of the larger crustaceans, a bright red lobster *Willemoesia* surfaced, snapping and splashing, to remind them who was absent from their work party and missing all the fun.

The most stunning catch of all in Valparaiso bay was the vertebrate fish. They had seen jack mackerel at the fish markets in Japan, but ten species here were entirely new. The famous anchovy—mainstay of the Chilean fishery—was out of season. Instead, silver meshes of sardines clogged the trawl.

Given its status as the richest fishing ground in the world, the trophic hierarchy of the Humboldt Current is surprisingly truncated. *Challenger*'s keel cleaved through carpets of red, green, and brown plankton, food for the hungry krill and copepods that, in turn, served marauding schools of *Sardinops sagax*. Mackerel and hake dominated the ranks of midsize fish, which nature reserved for the diet of whales and sea lions, but which modern-day supermarkets sell by the truckload as packaged fish fillets.

Since 2004, however, one tasty fillet staple, the Chilean hake (*Merluccius gayi*), has mostly disappeared from the shelves. *M. gayi* is a long, slender fish with a two-toned body, silver and blue, and an impressive array of sleek fins for rapid swimming. In *Challenger*'s time, it was a local seafood favorite preserved with salt, but the modern Peru-Chilean fishery has brought the hake population to the brink of collapse. Industrial trawl fishing selects for larger adult males, leaving a more diminutive, female-dominated population to cannibalize themselves to ensure the survival of spawning adults.

That said, climate change may be more decisive than overfishing in the decline of Chilean *Merluccius* because it threatens the coastal upwelling mechanics vital to the entire Humboldt Current ecosystem. The warming of surface waters over the last half century means a more stratified, sluggish

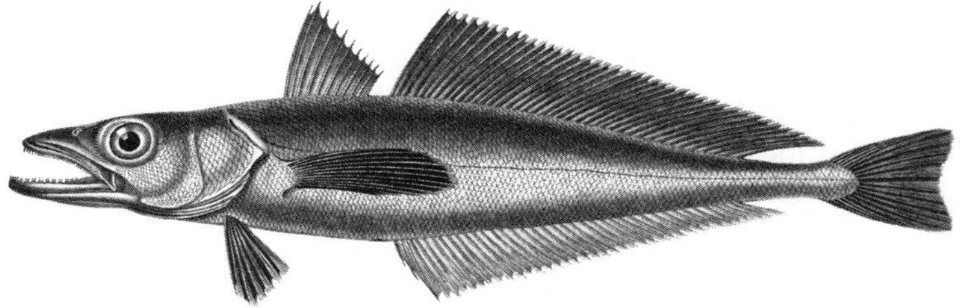

FIGURE 12.3 Once plentiful, stocks of Chilean hake (*Merluccius gayi*) face collapse due to overfishing. Claudio Gay, *Atlas de la Historia Física y Política de Chile*, vol. 2, plate 8 (1854).

ocean, with reduced mixing and less oxygen transfer between cold and warm layers. For the bottom-dwelling hake, fewer krill and copepods now survive the journey through the unventilated water column to provide fodder. If these challenges weren't enough, the voracious Humboldt squid (*Dosidicus gigas*) has taken advantage of overfished competition to expand its range southward, where it finds the silver-scaled hake much to its taste.

If *Merluccius* fails to survive, it won't be for want of trying. The hake's multiple adaptations to increased predatory pressures and food shortages have been spectacular—an example of the remarkable plasticity of species traits under environmental stress. *M. gayi* traditionally spawn in subsurface waters adjacent to the continental shelf; their larvae then migrate toward shore. But hake eggs are now smaller and laid closer to the coast compared to just a few decades ago, while juveniles have taken to congregating in remote places to avoid cannibalizing by their elders. The average hake is now barely half the size of its twentieth-century forebear and with a drastically reduced life cycle. Juveniles mature to reproductive adulthood in months rather than years. No sea creature better captures the retrospective idea of *Challenger*'s lost ocean or the reality of evolution in overdrive under climate change. The grilled hake you order this evening at a restaurant in Valparaiso might be the same species caught in *Challenger*'s trawl—but it is a very different fish.

Christmas was fast approaching, which meant long hours of rehearsal on deck for *Challenger*'s brass band and a much-anticipated holiday for the tired crew. The local hake has little cause to celebrate the season, however. Chilean fishermen historically associate Christmas with the arrival of the eponymous El Niño, an interannual influx of warm, depleted waters from the tropical north. El Niño conditions force *Merluccius* to spread out beyond its usual hunting grounds in search of food, while warmer temperatures exhaust the females, diminishing their fertility. Compounding the challenge, El Niño acts as a system reset for the entire Humboldt zone, allowing invasive species to establish fresh niches and put further competitive pressure on native fish. As the twenty-first century ushers in a less predictable, more intense El Niño regime for Pacific South America, hake populations will continue to fluctuate wildly—their natural response to an unnatural ocean—leaving coastal fisheries just as prone to collapse.

Like Tahiti and Juan Fernandez before it, Valparaiso was a scene of disappointment for the Challengers. They had looked forward to the majestic Andes, but the view was blocked by a line of scrubby red hills behind the town. Nothing about the European-styled settlement, treeless and dusty, suggested a vale of paradise. The chilled atmosphere produced by the Humboldt Current, combined with the perennial high pressure of the South Pacific zone, brings fog and clouds to mountainous Chile but without the balm of rain. The *Challenger* naturalists who set out on horseback to explore the hinterland soon found themselves, their horses, and all their gear caked in red dust. Six of the *Challenger* crew took the opportunity to desert, nevertheless.

In the second week of December, the branch of the coastal current south of Valparaiso floated them poleward but only into the teeth of persistent southerly gales. So they drifted to the west, once more in the company of leaping dolphins and a pod of whales, to the vicinity of the Juan Fernandez archipelago where they had recently been. Along their zigzag course, the officers tracked changes in the wind according to the behavior of their seabird retinue. The mollymawks, petrels, cape pigeons, and broad-winged albatrosses filled the skies above the flapping sails in a gale but inevitably vanished before a calm.

A Christmas dinner of roast beef, currant cakes, and satirical speeches relieved the mood of frustration in the officers' mess. The band played carols, followed by Captain Thomson on his cello. But festivities turned sour once the men broke into the grog stores. Rolling brawls broke out across the lower decks of the *Challenger*. One sailor named Williams had his jaw shattered in three places (official story: fell out of his hammock).

By New Year's Eve, they had reached their Pacific exit—the Messier Channel, gateway to the Strait of Magellan. After the howling maelstrom of their southward passage, the still silence of the Patagonian glades seemed alien. On both sides of the ship, a deep green forest rose steeply to a rocky summit. Beyond these, they could sense the loom of the massive southern glaciers crested in mist. The damp air crackled with cold. The contrast with arid, red Valparaiso was remarkable—like finding a hidden old world beyond the new. Lines of beech trees reached down to the water's edge, with overhanging rocks carpeted in mussels. The ship's artist, strolling onshore for a picturesque view, quickly found himself waist deep in a bank of moss.

A New Year's picnic in a fairy-like cove included delicious fillets of the local hake. Following *Challenger*'s own path, the southern cousin of the Chilean *Merluccius* leaves the titanic South Pacific currents for quiet Patagonian fjords, which it treats as a kind of nursery. The Challenger naturalists failed to distinguish between the two species, but *Merluccius australis*, its tender larvae feasting on copepods in placid waters, may represent the best chance of survival for the genus. Whatever happens in our near future ocean, desperate, fast-track adaptations to climate change by marine animals mean there will be no mistaking *M. australis* for *M. gayi* moving forward. What populations do not merge or die out will the more rapidly diverge in habitat and type, like objects in space that collide for an instant then spin away on opposite tracks through the cosmic wastes.

The Challenger philosophers, for their part, still believed in an ancient, mostly immutable ocean. Some species had modified over time, or came and went, but changes that did happen came at an unimaginably slow rate. Cocooned by that partial understanding, they enjoyed their New Year's hake dinner with easy spirits. After all, it was not hard to forget the upheavals of the oceanic past while reclining on a Patagonian rug of green moss and wildflowers. Grilled to a golden brown, the fresh-caught flesh of *M. australis* tasted creamy and succulent. As for the future, the weary Challengers looked only as far as the coming year—and home.

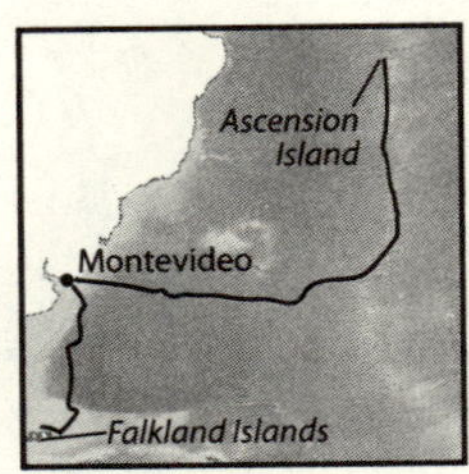

FIGURE 13.1 Green turtle (*Chelonia mydas*). Ramon de la Sagra, *Histoire physique, politique et naturelle de l'ile de Cuba: Reptiles*, plate 3 (1841). (Inset) HMS *Challenger* Leg 13. Falkland Islands to Ascension Island. February–March 1876.

CHAPTER 13

Dream of the Green Turtles

Ascension Island
7°56′ S, 14°22′ W

One rainy morning en route to Fiji in July 1874, a sharp cry of warning was heard from *Challenger*'s topmast. A short minute later, to the west, low-lying Turtle Island (Vatoa) appeared through the gloom, surrounded on all sides by a treacherous reef. Rocky corals in a palette of colors lurked beneath the water's surface as far as two miles from shore.

Islands named "Turtle Island" (or Isla Tortuga) are common among the world's tropical archipelagos, a sign of the sea turtle's historical ubiquity. A century prior, almost to the day, James Cook in *Resolution* had been the first European to chart this specific turtle haven southeast of Fiji, which he named for the green turtles (*Chelonia mydas*) that he observed swimming in numbers about the reef. Cook sent a boat to investigate. Approaching the beach, the *Resolution* party caught sight of a local crew fishing for *C. mydas*. They left trinkets on a rock as an invitation to parley, but the turtle hunters vanished into the woods beyond the beach.

The *Challenger* logs make no mention of Cook's green turtles, beloved in England for their healthful, succulent meat. But the Victorian explorers, though late on the scene, did not miss their opportunity entirely. Anchored in Galoa harbor, off Kadavu island in Fiji, the Challengers watched men from the local village load two bulky objects onto a double-hulled canoe (called a *drua*), then launch into the gentle surf, spreading a large triangular sail before them. The approach of this

FIGURE 13.2 Fijian catamaran, or *drua* (photograph by Frederick Hodgeson). *Scientific Results: Narrative*, vol. 1, plate 19 (1885).

prestigious craft caused enough excitement onboard to draw *Challenger*'s photographer on deck to record the moment. The officers' fondest hopes were soon realized. Drawing alongside, the Fijians announced themselves as an embassy from the local chief, with two precious green turtles as a goodwill gift to Captain Nares.

The Pacific green turtles—still alive and weighing more than two hundred pounds each—were carefully winched from the canoe to the *Challenger* deck where the captain's steward awaited them. Their shells, dried out from the sun, had lost their brilliant underwater sheen. The doomed animals now lay on their hard backs, too exhausted to move their flippers but still able to exhibit the traits of human-like despair for which their species was famous. A duet of heavy turtle sighs could be heard from across the deck. At closer range, one could detect the signature trail of tears from their bulbous, unblinking eyes.

Turtle meat had sustained oceangoing Europeans for centuries and, predating their arrival, held a totemic place in the food mythology of Pacific islanders. *Challenger*'s cook was thus well acquainted with the requirements for the officers' feast. First, he minced and cooked the turtle in the shell itself—a fatty, unctuous ragout—rendering fifty pounds of meat in all. Then, taking the great shell in hand—more olive-brown than green—he scraped strips of the glutinous cartilage from its interior. This *calipee* would serve as stock for a delicious turtle soup, renowned in Europe as the perfect stomach-settling appetizer to a grand banquet.

The chief's gift had a diplomatic significance not lost on Captain Nares. HMS *Challenger*'s career prior to her conversion to a research vessel had been as a naval patrol ship based at the Australian station. In August 1868, she had been sent to Fiji for the purpose of "punishing the hostile tribes" implicated in the death of a missionary. *Challenger*'s cannons, fired from offshore the island of Viti Levu, destroyed several villages, while parties of marines stormed the beach, guns blazing. "It is supposed that several of the natives were killed," reported the *Nautical Magazine* in the operation's aftermath. In fact, more than forty members of the Wainimala clan were killed. *Challenger* marines later raided and burnt another village whose inhabitants had refused to vacate their land in favor of a white settler. According to a report in the *Sydney Morning Herald*, the ensuing gun battle killed an unspecified number of Fijians.[1]

Challenger's deadly raids were fresh enough in local memory that, on the ship's return appearance off the coast in 1874, hundreds fled inland fearing renewed violence. In the harbor at Kandavu, Captain Nares and his officers were well aware that the presence of their ship aroused fear and loathing in their Fijian hosts. But both parties seemed committed to easing tensions. Colonizer and colonized could agree on one thing: the superior taste of *Chelonia mydas*. Flavored with sherry and generously spiced with nutmeg and cloves, turtle soup was a delicacy fit for a chief—or a distant queen.

Basic economics dictates that scarcity inflates value. The green turtle's high commodity status in the Victorian era emerges,

1 "Visit of H.M.S. Challenger to Fiji," *Sydney Morning Herald*, September 9, 1868. See also Jones (2023) 53.

accordingly, from a long history of unregulated slaughter and decline. Humans are, without doubt, the worst thing ever to happen to the world's sea turtles. Equipped with prehensile hands for gripping a turtle's shell, the hunter uses his lever-like arms to flip it over on its back—a simple, deadly trick no other predator can perform. A female green turtle might be fifty years old before she first becomes pregnant and looks to shore to nest. She then enjoys a two- to five-year hiatus before breeding again. With our beachcombing habits, however, she is pitifully vulnerable to capture during the brief hours of her long life that she spends laying eggs in the sand.

The human threat to turtles extends through their life cycle. Turtle eggs are prone to being dug up by humans and their invasive fellow travellers: rats, cats, dogs, and pigs. For those juveniles that survive, hunters armed with nets and spears pursue them to the coastal shallows where they graze among the seagrass. Turtles have long inspired human fascination as well as appetite. In addition to its evocative weeping and sighing, the turtle's dome-like shell suggests a little house, or even the patterned world of the first creation. But on tropical coasts and island beaches worldwide, turtle and human habitats have increasingly overlapped with fatal consequences for the oceans' most charismatic reptile.

Following eras of marine dominance dating to the Jurassic, the sea turtle first encountered its biological nemesis when Asian seafarers began to populate tropical Polynesia three millennia ago. Archaeological sites suggest an original turtle population in the millions. But within a few centuries of human settlement, *C. mydas* was in steep decline. Faced with the sudden scarcity of its prized protein source, Pacific Islanders instituted tight restrictions on turtle hunting—one of the first conservation measures in the historical record. Only elite hunters were permitted to target the green turtle, and turtle feasting was reserved for the chief and his retinue. Culinary focus likewise shifted from turtle meat to turtle eggs.

Turtle conservation, a product of environmental necessity, was elevated to sacred law by the work of mythmaking and ritual. At sites of early settlements in Polynesia, turtle remains are mixed indiscriminately with ordinary foods and domestic tools. With the sharp drop in turtle numbers, however, stark changes appear in the anthropological record. The turtle disappears from the common home, while elaborate protocols develop for turtle butchery and preparation, including the compulsory consumption of all edible parts. Discrete names are given to parts of the intestine and different types of fat. Turtle hunters are accompanied by a supervising priest. Taboo mandates that the people enjoy turtle dishes in ceremonial settings only, while the turtle itself assumes a totemic, god-like status. In Fiji, the manufacture of turtle nets is embedded in ritual practice and enriched by storytelling. During hunting season, women elders line the beach chanting songs inviting the turtles in from the sea. Illicit killing or eating of turtles is punishable by death.

A parallel mythology emerged in the Admiralty Islands north of Papua New Guinea, another historical turtle feeding ground. There, in March 1875, the *Challenger* naturalists observed turtle skulls hung alongside human skulls in a local chief's hut. Across western Pacific communities, the turtle's head was considered its most sacred body part, with turtle flesh ranked second only to human flesh in value. To the northeast, in the Marquesas, the killing of a turtle served as an "acceptable substitute for human sacrifice."[2]

This golden era of Pacific human-turtle coexistence did not last. With the establishment of European commercial ports in the early nineteenth century, indigenous islander protections of sea turtles rapidly crumbled. The global market demand for turtle meat prompted a hunting free-for-all for *C. mydas* (the fashion for turtle shell products likewise decimated its near relation, the hawksbill turtle). A transoceanic trade in live

2 See Allen (2007) 962.

green turtles ferried thousands to wealthy subscribers across Britain and to luxury vendors in London. Often, the buyer's name was written on the turtle's belly for priority shipping. By the time of *Challenger*'s arrival in West Polynesia, control of the millennia-old traffic in green turtles had passed definitively from island chiefs to the grandees and bureaucrats of the British Empire—a historic power shift recognized in the Fijian chief's gift to Captain Nares.

For all its reckless butchery, the Pacific market in green turtles paled in comparison to the Atlantic trade. In fact, *Challenger* had already visited the most active turtle killing fields in the world during its Caribbean sojourn in the spring of 1873. In the eyes of historians, the opening of the West Indies to plantation agriculture would not have been possible without *C. mydas*, which loomed in the European colonial imagination of the seventeenth century like a god-given bounty: a cheap, inexhaustible source of protein to feed themselves and their enslaved laborers abducted from West Africa. The entire Jamaican colony, for example, subsisted on turtle meat for almost two centuries. The largest turtle rookery in the world lay just to the south at the Cayman Islands, where a quasi-piratical industry sprang up to supply the ballooning caloric demands of the New World slave economy.

The impacts of West Indies colonization on the indigenous green turtle population beggar belief and have drawn comparisons to the near extinction of the millions-strong bison herds on the plains of the American West. Columbus in 1503 encountered a Caribbean Sea so full of turtles he was obliged to pause his ship to allow the migrating throngs to pass. A century and a half later, the Cayman Islands still boasted "infinnit numbers" of green turtles swarming its beaches in the summertime to lay their eggs. One estimate sets the pre-Columbian green turtle population of the West Indies at more than 90 million. Fast forward to today, perhaps a few hundred thousand Caribbean green turtles remain, a fraction of 1 percent of the original numbers.[3]

The first attempt to regulate the frenzied turtle harvest came as early as 1620, without success. By the late seventeenth century, some forty British sloops scoured the beaches of the Cayman Islands in the high season, each capturing up to fifty turtles a day. Later, steamships and tin packaging technology further bloated the transatlantic traffic and its profits. By the time of *Challenger*'s visit, the export market in live turtles from the West Indies and Key West had reached 15,000 per annum, in addition to 10,000 pounds of tinned "preserved" turtle delivered to British and American retailers.

The turtle death spiral was well underway by the 1870s despite the high-volume Caribbean trade that continued to flourish. The Bahamas had already been emptied of *C. mydas*. The Challengers leave no record of turtle sightings in their logs on the passage north from St. Thomas to Bermuda—a drastic ecosystem change since the time of Columbus. To the south, the Cayman Island turtlemen had extinguished the local population and turned their attention to the turtle rookeries of Costa Rica and Nicaragua three hundred miles away. Aided by the Miskito people—expert turtle hunters—the Cayman Islanders continued to supply the European and American markets for turtle products well into the twentieth century before those hunting grounds, too, were exhausted. Only in the 1970s, with the prospect of imminent extinction, did conservationists mobilize to save the legendary turtle. Governments across the tropical zone imposed bans on hunting turtles and harvesting their eggs—some two thousand years after the islanders of the South Pacific first regulated their turtle stocks.

Even in our age of mass extinction, green turtle losses since 1600 are staggering. For Columbus—and later European navigators William Dampier and Cook—"green-turtle geography"

3 See McClenachan et al. (2006); quoted in Parsons (1962) 28.

was central to their discovery experience of the tropical New World. They could not have believed in any threat to its ubiquitous existence. Incredible or not, however, the *Challenger*'s logs bear out the somber truth of its catastrophic demise over a few short centuries. It was not until the last months of their three-and-a-half-year global odyssey that *Challenger* finally encountered *C. mydas* in the wild. Nor had lessons been learned. The turtles of the remnant Atlantic population on remote Ascension Island had been earmarked for the stew pot or were languishing in artificial ponds awaiting shipment to British banquet tables.[4]

Wyville Thomson confided to his diary that the rainswept Falkland Islands were "not attractive and I doubt if they improve greatly on acquaintance." He spoke for the entire ship's company, impatient for home. A seaman named Thomas Bush drowned one night, apparently drunk, further souring the mood onboard. The rich kelp forests of Stanley Harbour, however, provided Thomson the opportunity to add to the expedition's bulging collection of sea stars and echinoderms—including a frilly, bright yellow sea slug (*Cladodactyla crocea*) whose mode of reproduction he pondered into the night. Northwards again to Montevideo, where they encountered their sister research vessel SMS *Gazelle*. The German government had been inspired by the *Challenger* expedition to mount its own round-the-world study of ocean circulation and biology (albeit on a smaller scale). At last, after ten days to rest and resupply, *Challenger* set out on the trackless wastes of the open Atlantic for the final time.[5]

From the beginning of the expedition, Thomson had authored regular contributions to the Royal Society *Transactions* and to the journal *Nature*, as well as short, popular

4 Parsons (1962) 11.
5 Thomson (1877) 2:203.

pieces for magazine publication. Now, a veritable tower of paperwork had built up on his desk in anticipation of their return. At Hawaii, a letter from the Admiralty had been waiting, requiring details of his plans for *Challenger*'s massive collections. During their monthslong transit of the Pacific, and now their final Atlantic leg, the prospect of publishing the expedition's results weighed increasingly on its director. It dawned on him that they had all radically underestimated the magnitude and complexity of the task, and that basic logistical questions were mounting: Where would he find the staff to process and analyze the thousands of jarred samples? To tabulate the ocean temperature data? How many years would it take to commission and author the dozens of reports? And would the government pay for it all as promised or—what was more likely—abandon the project before the work had been done to justify their long years at sea?

These professional pressures—lapping and overlapping in his mind like the ocean eddies at constant work on *Challenger*'s hull—meant Thomson missed their approach during the last week of March to Ascension Island, a volcanic cone rising dramatically from the sheer ocean midway between America and Africa. The island, in its peculiar isolation, stood directly in the path of the trade wind out of the southeast and was notorious for its surrounding rough seas. That morning, a violent squall had buffeted the ship, creating chaos in the workroom.

By the time Thomson emerged on deck the clouds had cleared, granting him a spectacular view of the distant island peaks rimmed by crimson light from the setting sun. At the foot of the mountains, the island lava beds, in changing shades of orange, red, and purple, undulated in the declining light as if still smoking from an eruption that happened yesterday. It was a raw, treeless vista—a moonscape dropped in the bright blue ocean. Except for a faint halo of vegetation visible at its highest peak, Ascension Island—Atlantic mecca for green turtles—appeared a burnt wasteland to human eyes.

The following day, with Murray and Moseley for company, Thomson embarked on the ascent of Green Mountain, the desert island's sole botanical attraction. The recent rainstorm had brought rare bursts of life to the barren ridges. Periwinkle beds in blue and white dotted the landscape among wild gooseberry bushes introduced from the African cape. But for the most part, it was a featureless climb until the last mile, which released them all of a sudden from the scorching lava desert into a forested oasis watered by ghostly fingers of mist playing about the summit.

On this sheer mountainside, on one of the world's most remote islands, it seemed they must have outrun the reach of their fellow human beings. But this tantalizing impression was false. Darwin's protégé Joseph Hooker had visited Ascension on board HMS *Erebus* in 1843. On his return to England, he had proposed to the Admiralty a bold plan for greening the island to ensure a reliable supply of spring water for visiting ships. Seedlings were duly imported from England and all corners of the Empire. A generation later, the Challengers contemplated the extraordinary fruits of that botanical experiment. Shaded glades of wattle trees and eucalypts from Australia buzzed with insect life. Mynah birds from India flew about their heads. As backdrop for their picnic lunch, an orchard and vegetable garden, alongside sheep grazing on the grassy slopes, seemed a landscape imported from a gentleman's estate in Kent.

Ascension's peak briefly restored for them a sense of the wild. It was strange to encounter land crabs scuttling about their feet at this elevation. A perilous track along the southeast cliff exposed them face-first to the blast of the trade wind and a plunging view of the sea below. From here they could take in the entire island. It was like a geography schoolbook illustration brought to life: a green band at the top, dark red and brown valleys in contoured relief below, and further out the blue, ever-restless ocean, flecked with surf, in unbroken visual rhythm with the panoramic sky.

Off the island's north shore rose a sheer rocky outcrop where a million sooty terns—called "wideawakes" for their grating call—kept their nests. On Ascension, great congregations of sea birds, spiralling upward like smoke, routinely blotted out the sun. Among them was the red-throated frigate bird, sworn nemesis of the island's newly hatched baby turtles on their desperate nighttime crawl from beach to sea.

Ascension Island's green turtles had acquired a new enemy, of course—one far deadlier than the frigate bird. The beach next to the Royal Navy garrison town on the bay was the chosen nesting place for thousands of *C. mydas* who made their transatlantic pilgrimage to Ascension between Christmas and midsummer. The steep slope ensured protection for nests at high tide, and the loose but firm sand provided the ideal platform for ocean-dwelling reptiles unused to walking. It was also perfect for nest digging. After years abroad, these turtles had swum a thousand miles from their feeding grounds off the Brazilian coast to reproduce on their own natal shore. Night after night, the females labored ashore while their male consorts cruised in the nearby surf awaiting their return.

But a nasty fate lay in store for the turtles, as the Challengers had ample opportunity to observe. Ascension Island was, by a bizarre bureaucratic fancy, designated a ship and "launched" in 1815 to supervise the captivity of Napoleon on St. Helena, 800 miles to the southeast. The economics of Ascension were simple: the garrison paid for the Admiralty's import of its supplies with the harvesting and sale of green turtles. When the moon rose, the sailors of the turtle watch took up their station on the margins of the rookery. Before long the first of the females came ashore, heaving and sighing. About a hundred feet up the beach she stopped, then with her front flipper began to dig vigorously. Settling herself, she deposited a hundred or more eggs in the hole she had made, each half the size of a billiard ball. By now, several others had joined her, attending their own nests at a discreet distance.

Though her own awkward track up the beach was obvious, the turtle took great pains to smooth over the sand where her eggs were buried. Added to her determination to lay eggs on the very beach where her mother had laid hers, she had inherited a deep fear of crabs, birds, and other hungry opportunists snatching her offspring (imported rats and feral cats now roamed Ascension). After two hours nest-building, the turtle turned at last to leave. The moment she stepped ponderously back down the beach in the approximate direction of Brazil, the sailors of the watch casually stood up. Walking over, they surrounded the animal and—with care to avoid her snapping mouth—gripped their hands under her shell. Then, in a single, coordinated motion, they flipped her over on her shell. To immobilize an animal weighing up to 800 pounds, teamwork was essential—and muscle power. Once they had repeated this maneuver for each of the thirty or forty turtles on the beach, they went back to bed.

In the morning, the captive turtles, exhausted by their ordeal, were winched into one of two bayside "ponds"—actually inground stone tanks—constructed for the purpose. The ponds were engineered to allow a continual flush of seawater from the bay but with sluices narrow enough to ensure no escape for the turtles. The pond managers kept juvenile turtles born from salvaged eggs as pets in buckets of seawater nearby. These they fed with cut-up raw meat.

Between the ponds stood a kind of gallows for stringing up the animals for slaughter. It was rarely idle. Turtle meat fed the garrison of several hundred men while every ship that stopped en route home from the East Indies expected their choice cuts of "Ascension beef" for the prevention of scurvy. Other ships arrived with the express purpose of transporting a bale of turtles alive to their buyers in England. These captive turtles suffered appalling conditions en route. At best, the turtle was packed in a barrel of water; at worst, it endured the entire journey upside down on deck, splashed with the occasional bucket. Diet on board consisted of scraps of meat and bread. Adult green turtles are herbivores. On an average voyage many turtles, perhaps most, perished.

Henry Moseley's interest in *Chelonia mydas* lay less with the turtles themselves than their eggs, which for some tropical island communities rated more highly as a delicacy than turtle meat. For Moseley, the eggs' most curious feature was a little wrinkle on the surface that did not fill out until full grown and hatching was imminent. To make best use of his brief time on Ascension, he had in mind an original study of the turtle embryo's development. So he returned to the Georgetown beach with a spade and a local guide. The sand the turtles had chosen was cool even in the noonday sun and moist to the touch. He excavated several nests that had been dug on different nights. Then he reburied each set of eggs in a bucket of sand and lugged these back to the ship.

Moseley's plan was to hatch the eggs artificially. He had read that turtle eggs responded to the warmth of the sun on their sandy nest, so he left the buckets on deck during the day then covered them over at night. Given his carefulness—and the turtles' supposed resilience—he was dismayed when every last egg died within a few days. From this failed experiment only a single, narrow conclusion could be ventured. Popular opinion of reproduction in *C. mydas* was not to be trusted. A turtle's eggs, in their natural setting, were laid at sufficient depth to maintain a constant, cool temperature for optimal growth in sand that never dried out from the sun. This explained the green turtle's devotion to her home shore in preference to all others. Her very existence, after all, was proof of the precisely congenial conditions to be found there.[6]

6 The preserved embryos from Moseley's unhatched eggs became the subject of a study published in the first of *Challenger*'s fifty volumes of scientific reports, see Parker (1880) 1:1–58.

Turtle embryos, like most reptiles, are uncommonly sensitive to temperature. Buried at a depth sufficient to insulate them from fluctuations in the day's temperatures, gradual warming of the egg during its two-month gestation derives from the metabolic energy of the embryo itself. Cool nighttime air serves as a trigger for hatching. The baby turtles then emerge en masse into the darkness, a strategy designed to increase their individual chances of survival against a phalanx of predators. Even a small increase in overnight temperatures—characteristic of current global warming—will botch the finely balanced process, leaving the embryo to rot in its egg.

Temperature during the egg's two-month gestation is likewise critical. Successful incubation of the green turtle embryo occurs only between 25°C (77°F) and 33°C (91°F). Since the visit of HMS *Challenger* 150 years ago, beach temperatures on Ascension Island have risen an estimated half degree, with more warming to come. Similar to other reptiles, a turtle embryo's sex is determined by temperatures during the middle period of incubation. The warmer the temperature, the greater the skewed proportion of females. A study from the early 2000s estimated the female turtle hatchlings of Ascension Island already at 75 percent of the total population. Another study—based on model projections—suggests the total feminization of green turtle communities worldwide by 2070 under an extreme climate change scenario: extinction, in other words.

But there is cause for tentative optimism. *C. mydas* has survived past global episodes of climate deterioration (if not at the current accelerated rates of change). And Ascension Island itself has proven a vital laboratory for study of the green turtle's natural genius for adaptation. In March 1876, Henry Moseley hitched a ride on a small naval steamboat deployed to collect turtles from Ascension's more than twenty rookeries. For a small island only seven miles across, the beaches he saw displayed a remarkable variety: some of black volcanic sand, others nearby of white sand. The green turtle eggs

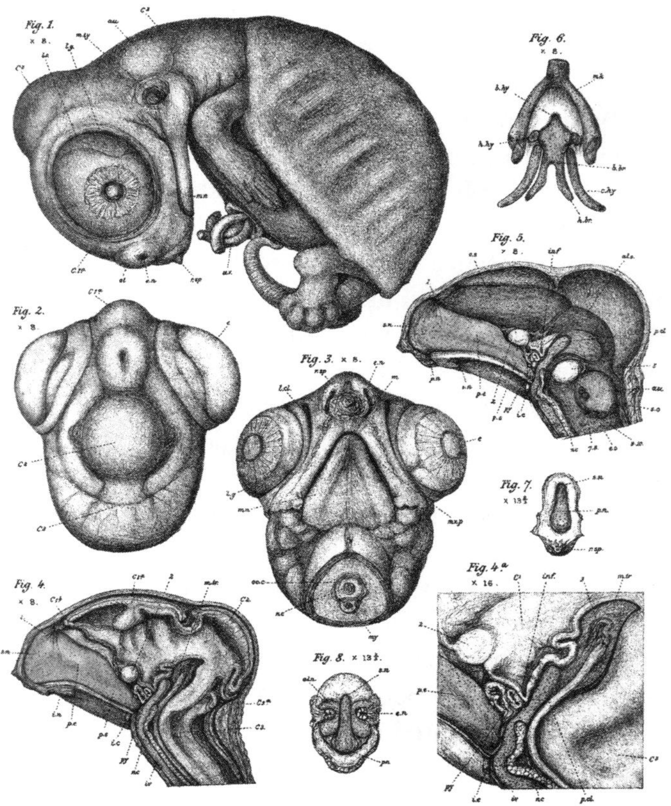

FIGURE 13.3 Green turtle embryo. *Scientific Results: Zoology*, vol. 1, plate 1 (1880).

nested on the volcanic beaches, it turns out, have adapted to the higher nesting temperatures of the sun-absorbing black sand. A turtle hatchling from one beach will have a measurably different temperature sensitivity to a turtle of the same species hatched on the beach next door. Prospects for *C. mydas* may seem bleak, but it would be wrong to underestimate their resilience.[7]

7 Black sand hatchlings will be predominantly female; white sand hatchlings, male.

After all, the green turtle has a well-established reputation for astonishing humans. Prior to Ascension's discovery by a Portuguese navigator in 1501, almost nothing was known about the turtle's ocean life and habits between their beachside birth and their return decades later to nest. As a volcanic pinnacle of land, Ascension Island has no coastal shelf. This deepwater shoreline means no seagrass or coral feeding grounds for the turtles. The existence of a major rookery a thousand miles from the turtle's natural habitat perplexed the philosophers of the age. Why travel so far? More baffling still, *how* did the creatures get there? As Charles Darwin put it in a letter to *Nature* a few months prior to *Challenger*'s departure, "How can we account for the turtles . . . finding their way to that speck of land in the midst of the great Atlantic Ocean?"[8]

The naturalists of HMS *Challenger* did not feel confident to speculate on this question, though many have since. Other long-range migrating species, such as birds, consult the map of the heavens—sun, moon, and stars—to orient their journey. But turtles are vulnerable to sharks if they linger too long at the surface, and have poor eyesight anyway. They are more likely to have imprinted the unique magnetic signature of their coastal destination. Earth's magnetic field consists of contour lines intersecting the surface at varying angles and intensity. To follow one such isoline between Ascension Island and Brazil might well represent the first life-or-death challenge for a baby turtle. The prevailing South Equatorial Current helps matters by bearing the turtle along, while the South American continent is, of course, a generous target. Most juvenile turtles arrive at a general area of the coast then turn north or south toward their favored neighborhood.

The turtle's return journey decades later to tiny Ascension—a forty-day swim against the current across a thousand miles of blank ocean—is an exponentially more difficult proposi-

8 Darwin (1873) 360.

tion and has been called "the most enigmatic masterpiece of long-distance navigation performed by any animal." Magnetic cues might not pinpoint the island precisely—also, these change over time. The turtle's likely solution is to rely on magnetic cues to bring them to the vicinity of the island, then to seek out wind-borne signals of its exact position. Rising to the surface to breathe, the swimming turtle's acute sense of smell detects rich odors of vegetation from Green Mountain and the reek of guano from the island's crowded tern rookeries. Diving down, the turtle uses the underwater pulse of waves breaking on the shore to navigate the final weary miles to its maternal beach.[9]

The turtles' mastery of magnetic navigation stands in stark contrast to their human tormentors. Like most discovery expeditions of the nineteenth century, HMS *Challenger*—at the Admiralty's behest—devoted vast funds and labor compiling geomagnetic data using the most sophisticated instruments of the age. The results never yielded the holy grail the explorers sought—a magnetic map of the world. Buried in those reams of data, however, were clues to a celebrated mystery of Victorian science: the Brazilian green turtles' wild odyssey to Ascension Island.

C. mydas has been a poster child for marine conservation since the 1960s—for good reason. The green turtle both symbolizes the historical exploitation of sea life by humans *and* embodies the full range of current threats posed by an Anthropocene ocean: rising seas inundate its nesting beaches, while warming temperatures compromise the viability of turtle hatchlings. Uncounted thousands of turtles annually fall victim to trawlers' nets. Plastic pollution, too, is an existential menace. Fishing debris traps and strangles the turtle, and transparent plastic trash that the turtle mistakes for a

9 Åkesson et al. (2001) 420.

gelatinous snack corrode its digestion. In a recent study, more than 50 percent of Ascension turtles at their feeding grounds in Brazil were found with plastic in their stomachs.

We share coastal habitats with the turtles but are obnoxious cotenants. An epidemic virus called *Fibropapillomatosis*—the consequence of shoreline algal blooms from toxic runoff—spawns bulky tumors on the turtle's body and flippers until it can't swim, see, or breathe. Marine heat waves—courtesy of human carbon emissions—depress turtle fertility. Even light pollution is a turtle problem. Baby green turtles hatched on Florida's populated beachfronts—programmed to distinguish between dark land and the relative brightness of the water—are fatally deceived by streetlights and head inland instead of toward the ocean. Adult turtles are famed for their navigation skills, but tens of thousands of turtle hatchlings die every year from simply turning the wrong way on the beach.

Perhaps the darkest cloud that looms for the green turtle is the threat to its food supply, the same faced by the celebrated manatees of Florida. Seagrasses—including *Thalassia*, known as "turtle grass"—have declined 30 percent globally since the 1870s from a variety of causes including subsurface warming, toxic runoff, and sedimentary buildup in coastal waters blocking out the life-giving sun. Seagrass die-off in turn creates a vacuum conducive to algal blooms and their resident turtle viruses such as *Fibropapillomatosis*, accelerating a deadly feedback loop.

That said, green turtles have not lost their power to surprise. Even in the face of this tsunami of challenges, populations of *C. mydas* are currently rebounding across the Atlantic. On Ascension Island alone—where the turtle has been protected since the 1950s—the number of turtle egg clutches laid annually in the last half century has increased sixfold to more than 23,000. Key beaches have increased their nesting numbers tenfold. The turtle has now recovered to 75 percent of its baseline population of 20,000 when the Royal Navy first colonized Ascension two centuries ago.

Similar data are reported at Aldabra Atoll, a major nesting ground in the Indian Ocean, while the recovery of the Hawaiian green turtle from near extinction "has occurred in a far shorter period of time than previously thought possible." In 2012, the International Union for Conservation of Nature downgraded the Hawaiian green turtle's status from endangered to "least concern." The secrets to this conservation success? Nothing more than basic measures strictly policed: a ban on turtle hunting and the harvesting of turtle eggs. Local conservationists and volunteers at breeding sites have likewise worked heroically to protect nests and increase turtle survival rates. The lesson of the green turtle's tentative recovery is that all is not lost. Protections work. Only better governance of the oceans is sorely needed—and the moral and technical will to shut off the human waste stream to the ocean.[10]

Indigenous tropical island communities have long intuited a deep connection between humans and turtles. Both are habitual migrants with international lifeways. In Polynesian mythology, turtles shape-shift as humans and are incarnated as gods. In one creation story, humans are born from eggs laid by the World Turtle. The *Challenger* voyage itself—crisscrossing oceans, bridging the land and deep sea—mirrored the turtle's odyssey. By these lights, the fate of the green turtles presages our own: the question remains which turtle storyline we choose to follow. Will adaptability and resilience be the tale of global humanity told by future historians? Or will we be reduced, like the post-Victorian sea turtle, to small bands of wanderers in a depleted world habitat—where *Challenger* sailed with barely a turtle in sight?

10 Balazs and Chaloupka (2004) 491.

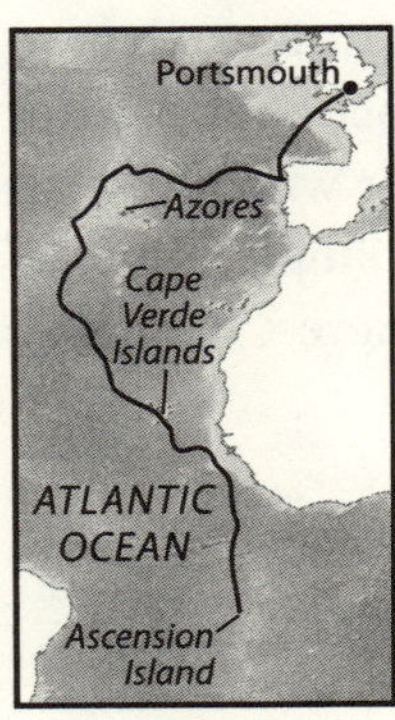

FIGURE E.1 HMS *Challenger* at Juan Fernandez. Frederick Whymper, *The Sea: Its Stirring Story of Adventure, Peril and Heroism* (1887). (Inset) HMS *Challenger* Leg 14. Ascension Island to Spithead. April–May 1876.

EPILOGUE

On April 3, 1876, at first light, *Challenger* set sail out of Clarence Bay. By noon, the unearthly peaks of Ascension Island had slipped from view beyond the stern. With their bellies full of lip-smacking turtle broth, the Challengers pursued a northeasterly course directly opposite the enigmatic chelonian migration but with a similar goal: faraway home. Taking full advantage of a stiff trade wind out of the southeast, they crossed the Equator for the sixth and last time four days later. Day by day, as the shores of England approached, the scientific staff experienced alternating waves of relief and apprehension. On their final passage to Portsmouth, unseasonable winds out of the northeast slowed them to a crawl. Provisions—and good humor—were in short supply. "In our weariness and impatience," Thomson wrote, "we were driven to the verge of despair."[1]

For three and a half years, they had watched from the ship's deck while great whales sported alongside schools of dolphins and porpoises. At closer quarters, in their boats, they had observed fish of all kinds: silvery, with fins like fans, armed with teeth, or dotted with luminous spots. Languid jellies and cuttlefish abounded throughout the voyage, as did the miniature *Velella* and *Physalia*, spreading their tiny sails to the wind. From their tow nets, they collected original mollusks, crustaceans, and worms with their wriggling larvae, and scooped up clouds of microscopic plankton from the sunstruck surface of the waters. And from the deep sea, with the dredge and

1 Thomson (1877) 2:270.

trawl, they had raised an entire novel cornucopia of marine life—angels and monsters alike. But the philosophers of HMS *Challenger* were fed up at last with the business of exploring the oceans. They knew too that the real work of the expedition lay ahead of them in the laborious writing up of reports. The *Challenger*'s fame depended on it. Doubtful weather loomed beyond the longed-for horizon.

No public statues commemorate Wyville Thomson, with his unfashionable competence and more unfashionable ideas. *Challenger*'s fifty volumes of scientific reports are his monument. With his controversial decision to commission the expedition reports from experts regardless of nationality, Thomson helped foster an international community of marine researchers that is one of *Challenger*'s greatest legacies. No fewer than 76 authors, from six nations, contributed to the expedition reports. Generous as he was with sharing scholarly credit, Thomson fully intended to reserve his favorite phyla—the spiny echinoderms—for himself. That is, until the *Challenger* curse struck. "The strain both mental and physical was long and severe," he admitted on their return, "and it has told a good deal upon all of us." Four years later he was dead, buried by work, with the essay on echinoderms unwritten. He died a victim of his own success, overwhelmed by the sheer volume and variety of *Challenger*'s great haul.[2]

From the outset of the mission, Thomson had expressed contempt for parochialism in science and looked abroad. Germany zoology, for instance, was well established at the universities with teaching posts aplenty—unlike in Britain. For a generation of ambitious German students attracted by the glamour of evolutionary theory, Darwin's *Origin* had opened up a near infinite set of research questions for which the oceans were the richest laboratory. Well aware of developments on the continent, Thomson had been careful to recruit a German and

2 Thomson (1877) 1:xx.

Swiss-German to his *Challenger* staff, in addition to Henry Moseley, who had studied in Leipzig.

On *Challenger*'s return, the government predictably strangled funding for publishing its results before they were remotely done. Then Wyville Thomson died. At that dire juncture, the entire international effort might have collapsed under its own weight but for the intervention of an underachieving colleague Thomson had recruited at the last minute: John Murray. In addition to his Olympian administrative skills, Murray helped fund long-term staff and resources at "Challenger House" on Queen Street in Edinburgh, which churned out groundbreaking research reports for two full decades.

It bears repeating that Wyville Thomson, whom Murray replaced on his untimely death, deserves a higher standing in public memory. As *Challenger*'s scientific director, he supervised a breathtaking array of achievements in ocean research, both aided and inhibited by his skeptical turn of mind. Thomson was, at heart, an aesthete, a taxonomist best suited to the miniature scales of nature—but he lived and worked in an era of grand theories of life. He rejected his friend William Carpenter's "magnificent" model of general ocean circulation and, though he endorsed a form of Darwinian evolution, he could not commit to competitive selection as its mechanism. We can imagine that Thomson—as the public face of ocean science on *Challenger*'s return—felt pressure to solve the Great Question of the age: to explain how things faded in and out of being. But such a comprehensive vision was beyond his grasp. The beauty of animal forms under the microscope invariably distracted him just as he felt on the verge of some global insight.

But if Thomson was on the wrong side of main currents of Victorian intellectual history, he was strikingly ahead of his time in other ways. Deeply Forbesian in outlook, he focused on what he and Carpenter called "submarine climate"—changes in ocean temperature over time—as the engine of evolution. According to Thomson's idiosyncratic variant on Darwinism, environmental conditions—ocean currents, temperature, chemistry, and biota—determined the origin of species, and their extinction.[3]

Thomson's brave new ecology of the oceans shaped *Challenger*'s research program and is a principal reason the expedition looms large to marine scientists in the age of climate change. For Darwin, species evolution occurred in a nondirectional world—the "uniformitarian" model of natural cycles inspired by his mentor, geologist Charles Lyell. For Wyville Thomson and his Challenger School, by contrast, evolution takes place in a directional world of changing climate conditions. Of the two, Thomson's philosophy is more directly applicable to our current ocean—the ocean of the Anthropocene. We inhabit, in the twenty-first century, a directional world driven by climate change and ocean pollution, not a pure Darwinian biosphere where natural selection operates against a backdrop of stable (or naturally variable) environmental conditions. With Wyville Thomson at their head, the philosophers of HMS *Challenger* are sailing into their moment.

The past, wrote novelist L. P. Hartley, is a foreign country. For Earth's animals and their habitats, the past is more like a foreign planet. This is especially true of the world's oceans, which have absorbed the vast majority of extra carbon emissions since industrialization. The current velocity of warming—the rate at which isotherms shift across space—is up to seven times higher on ocean than on land.

Marine life is mortally sensitive to changes in ocean temperature and chemistry. A 2012 study of rising sea temperatures that compared *Challenger* data to current temperatures found that "the modern upper ocean is substantially warmer than the ocean . . . in the 1870s and that the warming signal is global in extent." Warmer water temperatures demand

3 Carpenter, Jeffreys, and Thomson (1869–1870) 474.

higher metabolic performance from all organisms that, in turn, pushes the envelope of oxygen availability. When critical oxygen thresholds are crossed, populations crash. Vulnerability is increased by the complex life cycles of many species, which involve a dispersive larval stage intersecting with multiple habitats and temperature regimes. Tolerance levels risk being exceeded at any phase.[4]

Biodiversity is the first casualty of a warmer ocean. More acidic seas threaten coral reefs and the multitude of core marine species who build shells. Further up the food chain, industrial fisheries deplete key predators such as tuna, shrimp, and squid, whose tasty meat now contains measurable amounts of toxins from mercury (a waste product of coal burning) to microplastics. Plastic, of course, is just the most conspicuous of human waste products now cluttering the oceans. Invisible killers—toxic runoff, radioactive waste, munition dumps, abandoned fishing gear, and raw sewage—have been accumulating for generations, with impacts on the marine habitat that can only be imagined since they have never been properly measured.

Global ocean biodiversity, prior to the Anthropocene, had been stable for millions of years, with species distributions settled since well before the beginning of the Pleistocene epoch. Under the current global warming regime, ocean habitats are shifting toward the poles at an average rate of 45 miles (72 kilometers) per decade. With this increasing tropicalization of temperate zones, species with narrow thermal tolerance for reproduction are forced to migrate or face extinction. During the most drastic ocean extinction event in Earth's history—the end-Permian warming 251 million years ago—marine animals simply "ran out of climatically habitable space." The current mass extinction—the sixth—exhibits the same pattern. Fish that survive will be smaller in size and

4 Roemmich et al. (2012) 427.

number. A few accidental winners, like jellyfish, will thrive, but marine biodiversity overall will decline precipitously—accelerated by relentless overfishing. During the turmoil, the abyssal deeps plumbed by HMS *Challenger* will be a refuge for ocean life and biodiversity as it has in extinctions past, until conditions there too pass beyond the pale.[5]

It is said that to heal a broken heart requires half the length of time the love affair itself lasted. On that principle, we might imagine our impact on the world's oceans—enacted over a century or so—will take mere decades to dissipate and the oceans return to their prior state. Not so. The actual timetable for the full recovery of ocean biodiversity—according to end-Permian precedent—will be between 10 and 40 million years. Even by this distant reckoning, the Victorian oceans the *Challenger* sailed are lost forever. Just as the end-Permian event saw the demise of brachiopods and crinoids and the rise of corals and mollusks to take their place, changes in species composition will make the post-Anthropocene ocean different from any yet seen on Earth. Today's rapidly changing marine environment is a mere preview of the main act to come.[6]

Which returns us to the *Challenger* paradox. The legendary expedition, by one measure, represents the vanguard of twentieth-century industrialization of the oceans, even of climate change itself. The Challengers pioneered forms of deep-sea sounding, mapping, and trawling that evolved into core technologies of the oil drilling and open-sea fishing industries. By opening the deep sea to science, the *Challenger* naturalists likewise opened new frontiers for resource extraction and profit, including deep-sea mining. Then there's the human, historical account to be totaled up. HMS *Challenger*—which had earlier served as a police vessel in the Australasian Pacific—relied on a colonial network of ports and supply lines

5 Pinsky et al. (2020) 154.
6 Alroy (2008).

to facilitate its journey of discovery. Enslaved and indigenous workers en route provided the labor and, by extension, the infrastructure for every last mile she sailed. The *Challenger* expedition, by this reckoning, is part of the great imperial stain on the nineteenth century.

By an alternate measure, however, *Challenger*'s place in what the poet John Keats called the "grand march of intellect" is secure. For modern oceanographers, Thomson's expedition is their Book of Genesis. *Challenger*'s soundings, at over five hundred stations worldwide, produced not simply a blizzard of data but a comprehensive vertical temperature profile for the world's ocean basins. Sediment samples from these soundings likewise produced the first profile map of the ocean floor, identifying critical trenches and ridges that would ultimately give rise to the defining Earth science breakthrough of the twentieth century: plate tectonics.

As for marine life, the fifty volumes of *Challenger* reports identify and describe an entire new zoology of the deep sea. To cite just one statistic: before 1872, no more than thirty species of deep-sea fish were known; *Challenger* raised that number to more than three hundred. The total of bottled specimens crated and sent home exceeded a hundred thousand. These and the voluminous temperature and water chemistry data inspired hundreds of research papers in the ensuing decades, effectively launching the modern marine sciences. John Murray himself coined the word "oceanography," the new field *Challenger* brought into being. Seen more broadly, the expedition marked, in Murray's own partisan estimation, "the greatest advance in the knowledge of our planet since the . . . fifteenth and sixteenth centuries."[7]

The fact the *Challenger* mission—in its ambition and scope—has never been emulated is one sure indicator of its historical importance. *Challenger*'s faithful, daily accounting

7 *Scientific Results. Summary* (1895) 1:xii.

of species distributions and sea temperatures worldwide for three and a half years in the mid-1870s stands as a unique glimpse into the marine environment before modern industrial fishing, waste pollution, and climate change drastically altered the character of our oceans. *Challenger*'s unanticipated legacy, with other expeditions of the eighteenth and nineteenth centuries, is thus as a time capsule or "vault" of real-world observations that collectively establish a baseline by which to measure the ocean's transformation in the Anthropocene. Without the *Challenger* voyage as that baseline, the modern ocean conservation movement could not be imagined.[8]

Historical baselines are, of course, relative. As we have seen throughout this book, *Challenger*'s oceans were no longer fully wild—altered in many respects from the precolonial period. Cod stocks and oyster beds had been scoured, and whale, seal, and turtle populations hunted to near extinction. But these human impacts are dwarfed, by orders of magnitude, by the strip-mining of ocean habitats in the 150 years since *Challenger*. Between their oceans and ours lies an unprecedented biotic upheaval we are only just beginning to comprehend.

HMS *Challenger* was aptly named. The epic voyage challenged Victorian indifference to the deep sea, as it does ours today. "A flood of light," announced John Murray in a public lecture he gave on *Challenger*'s return, "has been thrown on a vast region of the earth's surface about which before all was doubt, guess-work and ignorance." Without the work of *Challenger* the majestic diversity of the life aquatic could not have swum into our ken. Without the example of a multiyear research enterprise devoted exclusively to their study, the oceans could not have become an object of knowledge—and concern. The massive resources and labor invested in *Challenger*'s epic voyage, followed by the decades-long international effort to publish its results, speaks to the emergence in the late nine-

8 See Lister et al. (2011); K. Johnson et al. (2011); and Rillo et al. (2019).

teenth century of a new attitude toward the ocean as valuable *in itself.* The *Challenger* paradox, in this sense, might be understood as a form of totemism: we venerate what we destroy, as did the original hunters.[9]

To the philosophers of HMS *Challenger* themselves, the experience was more mundane, even profoundly tedious for long periods. It seemed they spent entire years—a minor lifetime—adrift on the reflective membrane of the sea's surface, leaning out over the ship's rail waiting for the dredge to appear. As one day passed indistinguishably into the next, the waiting itself became a philosophical act, since it invited contemplation. During a long voyage at sea, time crystallizes. Looking out over *Challenger*'s frothy wake was like looking at the literal past since the churning water carried the outline of the ship's hull still on it. In the end they sailed 70,000 miles in this monotonous style, with the past a trailing addendum to the present and the future an ever-receding line where sea met sky.

The voyage of HMS *Challenger* was, in essence, an experiment—a wooden cocoon for a new kind of experience. The emerging Victorian discipline of professional science—replacing religion—required that its shipboard philosophers renounce human concerns and contemplate the hidden realm of nature's beginnings: the original world of saltwater invertebrates, what most life was and is. The naturalists, officers, and crew of HMS *Challenger* are long gone, but the ship's imprint is still on the world's oceans. *Challenger*'s legacy—all fifty volumes of it—is a Victorian bulwark against our all-consuming present, her hold brimful with clammy wonders and lessons for the future. On a watery sphere like ours, the voyage never really ends. There she is still, the ship of legend, tripping the wind on the oceans of memory.

9 Murray (1877) 139.

ACKNOWLEDGMENTS

Thanks first of all are due to the Carnegie Foundation for their generous support of my *Challenger* research, comprising this book and an expedition database hosted on the Scottish Association for Marine Science (SAMS) website. For patient assistance during my time spent at the Natural History Museum in London, sincere thanks to librarians Lisa Cardy and Andrea Hart; also to Lauren Hughes for her expert behind-the-scenes tour of the *Challenger* collections; and to Giles Miller for conversations about them. The encouragement of Nick Owens, director of SAMS (and John Murray custodian), has been vital to the realization of this project in its many dimensions. He and Anuschka Miller welcomed this itinerant scholar from land-locked central Illinois to maritime Scotland with open arms. I'm grateful, too, to the UK Challenger Society for accommodating a historian of science at their biennial assemblies. My experiences at the conferences in London and Oban were highlights of my research experience and introduced me to a transatlantic network of *Challenger* scholars, educators, and boosters, including Erika Jones at the National Maritime Museum, Peter Tuddenham and Tina Bishop of the College of Exploration, and Philip Pearson. For expert suggestions to improve the book manuscript, I'm indebted to Anton Edwards and four anonymous reviewers at Princeton University Press. At the press, Ingrid Gnerlich, as always, offered unwavering confidence and gentle advice where needed, while Whitney Rauenhorst steered production of the book with cheerful professionalism. At the University of Illinois, the Research Board provided much-needed support for the project in its early stages. And on the home front, Nancy tolerated frequent research absences and the intrusion of dripping scuba equipment into the living room. Some debts can never be repaid.

BIBLIOGRAPHY

The fifty volumes of published results from the Challenger expedition are officially cited as follows (and referred to hereafter as *Scientific Results*):

Report of the Scientific Results of the Voyage of H.M.S. Challenger during the years 1872–76, under the command of Captain Sir George S. Nares and the late Captain Frank Tourle Thomson, prepared under the superintendence of the late Sir. C. Wyville Thomson, and now of John Murray, one of the naturalists of the expedition. 1880–1895. 50 vols. London.

GENERAL REFERENCE

Abulafia, David. 2019. *The Boundless Sea: A Human History of the Oceans.* Oxford: Oxford University Press.

Adler, Antony. 2019. *Neptune's Laboratory: Fantasy, Fear, and Science at Sea.* Cambridge, MA: Harvard University Press.

——. 2023. "Changing Narratives in the History of Oceanography," in *Handbook of the Historiography of the Earth and Environmental Sciences*, ed. Elena Aronova, David Sepkoski, and Marco Tamborini. New York: Springer.

Carson, Rachel. 1951. *The Sea Around Us.* Oxford: Oxford University Press.

Cohen, Margaret, ed. 2021. *A Cultural History of the Sea in the Age of Empire.* London: Bloomsbury Academic.

Denny, Mark. 2008. *How the Ocean Works: An Introduction to Oceanography.* Princeton: Princeton University Press.

Duffy, J. Emmett. 2021. *Ocean Ecology: Marine Life in the Age of Humans.* Princeton: Princeton University Press.

First Global Integrated Marine Assessment: United Nations. 2016. Cambridge: Cambridge University Press.

Herring, Peter. 2002. *The Biology of the Deep Ocean.* Oxford: Oxford University Press.

Jones, Ian, and Joyce Jones. 1992. *Oceanography in the Days of Sail.* Sydney: Hale & Iremonger.

Konishi, Shino, et al., eds. 2015. *Indigenous Intermediaries: New Perspectives on Exploration Archives.* Canberra: ANU Press.

Levinton, Jeffrey S. 2014. *Marine Biology: Function, Biodiversity, Ecology.* 4th ed. Oxford: Oxford University Press.

Mills, Eric L. 2009. *The Fluid Envelope of Our Planet: How the Study of Ocean Currents Became a Science.* Toronto: University of Toronto Press.

Reidy, Michael S., and Helen M. Rozwadowski. 2014. "The Spaces in Between: Science, Ocean, Empire." *Isis* 105:338–51.

Roberts, Callum. 2007. *The Unnatural History of the Sea*. London: Island Press.

Rohling, Eelco J. 2017. *The Oceans: A Deep History*. Princeton: Princeton University Press.

Rozwadowski, Helen. 2018. *Vast Expanses: A History of the Oceans*. London: Reaktion Books.

Ruppert, Edward E., Richard S. Fox, and Robert D. Barnes. 2004. *Invertebrate Zoology*. 7th ed. Andover, UK: Cengage Learning.

Special Report on the Ocean and Cryosphere in a Changing Climate. 2019. Intergovernmental Panel on Climate Change.

PRIMARY SOURCES: UNPUBLISHED MANUSCRIPTS

Books on the *Challenger* voyage have relied mostly on the official narrative and the published accounts of Thomson, Moseley, Swire, Spry, Campbell, Wild, and Matkin. To offer a fresh account, with novel details of life aboard, I have drawn substantially on the ship's logs as well as unpublished diaries of the voyage, two of which are previously uncited in the *Challenger* literature. The first was kept by Lieutenant Bromley, the other (unsigned) by an officer whose identity remains elusive—copies of both are to be found at London's Natural History Museum library. In addition, I have consulted the unpublished journals of Thomson, Murray, and Willemoes-Suhm, also in the Natural History Museum archives. Much of this material was never revised for publication and has scant presence in expedition histories.

Aldrich, Pelham. *Journal*. Royal Geographical Society. RGS213270.

[Anon.]. *Journal*, H.M.S. *Challenger*, November 1872–June 1876. MS 10264. State Library of Victoria, Melbourne [copy at Natural History Museum, London].

Balfour, Andrew F. *Journal*. Royal Geographical Society. RGS213281.

Bromley, Arthur C. B. *Logbook* [Journal]. Murray Collection MSS BRO. Natural History Museum, London.

Channer, A. *Sketches and Journal*. Murray Collection MSS CHAN. Natural History Museum, London.

HMS *Challenger*: Ship's Log, 1872–1876. Murray Collection, Section 1,18–20. Natural History Museum, London.

Murray, Sir John. *Diary*. 5 vols. Murray Collection, Section 1,1–5. Natural History Museum, London.

Willemoes-Suhm, Rudolf von. *Diary*, 1872–1875. Murray Collection, Section 1,12. Natural History Museum, London.

Thomson, Sir Charles Wyville. *Diary*. Murray Collection, Section 1,10. Natural History Museum, London.

PRIMARY SOURCES: PUBLISHED

Brady, Henry Bowman. 1884. "Report on the Foraminifera." *Scientific Results: Zoology* 9:1–752.

Buchan, Alexander. 1889. "Report on Atmospheric Circulation." *Scientific Results. Physics and Chemistry* 2:1–75.

——. 1895. "Report on Oceanic Circulation." *Scientific Results. Summary* 2:1–38.

Buchanan, John Young. 1876. "Preliminary Report to Professor Wyville Thomson . . . on Work (Chemical and Geological) done on board H.M.S. Challenger." *Proceedings of the Royal Society* 24:593–623.

——. 1884. "Report on the Specific Gravity of Samples of Ocean Water." *Scientific Results. Physics and Chemistry* 1:1–46.

——. 1886. "On Similarities in the Physical Geography of the Great Oceans." *Proceedings of the Royal Geographical Society* 8:753–70.

——. 1887a. "On Ice and Brines." *Proceedings of the Royal Society of Edinburgh* 14:129–47.

——. 1887b. "On the Distribution of Temperature in the Antarctic Ocean." *Proceedings of the Royal Society of Edinburgh* 14:147–49.

——. 1913. *Scientific Papers*. Vol. 1. Cambridge: Cambridge University Press.

——. 1919. *Accounts Rendered of Work Done and Things Seen*. Cambridge: Cambridge University Press.

Campbell, Lord George. 1877. *Log Letters from* The Challenger. London: Macmillan.

Carpenter, William B. 1870. "On the Temperature and Animal Life of the Deep Sea." *Nature* (March 10):488–90; (March 24):540–42; (March 31):563–65.

——. 1870–1871. "On the Gibraltar Current, the Gulf Stream, and the General Oceanic Circulation." *Proceedings of the Royal Geographical Society* 15:54–91.

Carpenter, William B., Gwyn Jeffreys, and Wyville Thomson. 1869–1870. "Preliminary Report of the Scientific Exploration of the Deep Sea in H.M. Surveying Vessel 'Porcupine' during the Summer of 1869." *Proceedings of the Royal Society* 18:416–80.

Carpenter, William B., and Wyville Thomson. 1869. "Preliminary Report of Dredging Operations in the Seas to the North of the British Islands, in Her Majesty's Steam Vessel *Lightning*." *Proceedings of the Royal Society* 17:168–200.

Darwin, Charles. 1859. *On the Origin of Species*. London: John Murray.

Dittmar, William. 1895. "Report on the Composition of Ocean Water." *Scientific Results. Physics and Chemistry* 1:2–251.

Günther, Albert. 1880. "Report on the Shore Fishes." *Scientific Results: Zoology* 1:1–82.

——. 1881. "Report on the Deep Sea Fishes." *Scientific Results: Zoology* 22:1–329.

Haeckel, Ernst. 1882. "Report on the Deep Sea Medusae." *Scientific Results: Zoology* 4:i–cv; 1–154.

——. 1887a. *The History of Creation*, trans. Ray Lankester. 2 vols. New York: Appleton & Co.

——. 1887b. "Report on the Radiolaria." *Scientific Results: Zoology* 18.1–2: i–clxxxviii; 1–1762.

——. 1888. "Report on the Siphonophorae." *Scientific Results: Zoology* 28:1–380.

Hoek, P.P.C. 1883. "Report on the Cirripedia." *Scientific Results: Zoology* 8:1–186.

Hooker, Joseph Dalton. 1877. "President's Address." *Proceedings of the Royal Society of London* 25.171–78:339–62.

Hoyle, William Evans. 1886. "Report on the Cephalopoda." *Scientific Results: Zoology* 16:1–236.

Huxley, Thomas H. 1880. "The First Volume of the Publications of the *Challenger*." *Nature* 23:1–3.

"John Young Buchanan." 1926. *Proceedings of the Royal Society of Edinburgh* 45:364–67.

Lendenfeld, Robert von. 1887. "Report on the Structure of the Phosphorescent Organs of Fishes." *Scientific Results: Zoology* 22:277–329.

Lyman, Theodore. 1882. "Report on the Ophiuroidea." *Scientific Results: Zoology* 5:1–368.

[Matkin, Joseph]. 1992. *At Sea with the Scientifics: The* Challenger *Letters of Joseph Matkin*. Ed. Philip F. Rehbock. Honolulu: University of Hawaii Press.

Maury, Matthew. 1857. *The Physical Geography of the Sea*. New York: Harper Bros.

McIntosh, William C. 1887. "Report on the Annelida Polychaeta." *Scientific Results: Zoology* 12:1–532.

Moseley, H. N. 1876a. "On the Structure and Relations of the Alcyonarian *Heliopora caerula*." *Philosophical Transactions of the Royal Society* 166:91–129.

———. 1876b. "On the Structure of a Species of *Millepora* Occuring at Tahiti." *Proceedings of the Royal Society* 24:448–50.

———. 1876c. "Preliminary Report . . . on the True Corals Dredged by H.M.S. 'Challenger' in Deep Water." *Proceedings of the Royal Society* 24:544–69.

———. 1876–1877. "Preliminary Note on the Structure of the Stylasteridae." *Proceedings of the Royal Society of London* 25:93–101.

———. 1877. "On the Colouring Matters of Various Animals, and Especially of Deep-Sea Forms Dredged by H.M.S. Challenger." *Quarterly Journal of Microscopical Science* 17:1–23.

———. 1880. "Deep-Sea Dredging and Life in the Deep Sea." *Nature* 21:543–47, 569–93.

———. 1881. "Certain Hydroid, Alcyonarian and Madreporian Corals." *Scientific Results: Zoology* 2:1–208.

———. 1883. "On the Microscopic Characters of Volcanic Ashes and Cosmic Dust, and Their Distribution in Deep Sea Deposits." *Proceedings of the Royal Society of Edinburgh* 12:474–95.

———. 1884. "Address [on Pelagic and Deep-sea Life]." *Report of the British Association for the Advancement of Science* 54:746–53.

———. 1885. "The Fauna of the Sea-Shore." *Popular Science Monthly* 27:623–25.

———. 1887. "Report on the Structure of the Peculiar Organs on the Head of *Ipnops*." *Scientific Results: Zoology* 22:267–76.

———. 1892. *Notes by a Naturalist: An Account of Observations Made during the Voyage of H.M.S. Challenger Round the World in the Years 1872–1876*. 2nd ed. London: John Murray.

Murray, John. 1876. "Preliminary Report on Specimens of the Sea-bottoms obtained . . . [by] H.M.S. *Challenger*." *Proceedings of the Royal Society* 24:471–544.

———. 1877. "The Cruise of the *Challenger*." *Science Lectures for the People Delivered in Manchester*. Manchester: J. Heywood.

———. 1878. "On the Distribution of Volcanic Debris over the Floor of the Ocean." *Proceedings of the Royal Society of Edinburgh* 9:247–61.

———. 1879–1880. "On the Structure and Origin of Coral Reefs and Islands." *Proceedings of the Royal Society of Edinburgh* 10:505–18.

——. 1889–1890. "On Coral Reefs and Other Carbonate of Lime Formations in Modern Seas." *Proceedings of the Royal Society of Edinburgh* 20:79–109.

——. 1895. *Summary of the Scientific Results* [HMS *Challenger* Reports].

——. 1896. *On the Deep and Shallow Water Marine Fauna of the Kerguelen Region of the Great Southern Ocean*. Edinburgh: Robert Grant & Son.

Murray, John, and A. F. Renard. 1883. "On the Microscopic Characters of Volcanic Ashes and Cosmic Dust, and their Distribution in the Deep Sea Deposits." *Proceedings of the Royal Society of Edinburgh* 12:474–95.

——. 1891. "Report on Deep Sea Deposits." *Scientific Results. Deep Sea Deposits* 1–406.

Parker, William K. 1880. "Report on the Development of the Green Turtle." *Scientific Results: Zoology* 1:1–58.

Quelch, John J. 1886. "Report on the Reef Corals." *Scientific Results: Zoology* 16:1–194.

Redfern, Peter. 1888. "Sir Charles Wyville Thomson." *Proceedings of the Royal Society of Edinburgh* 14:58–80.

"Report on the Deep-Sea Temperatures Obtained by the Officers of *H.M.S. Challenger*." *Scientific Results. Physics and Chemistry* 1.

Sars, G. O. 1885. "Report on the Schizopoda." *Scientific Results: Zoology* 13:1–228.

Schulze, F. E. 1881. "Report on the Hexactinellida." *Scientific Results: Zoology* 21:1–504.

Smith, Edgar A. 1885. "Report on the Lamellibranchiata." *Scientific Results: Zoology* 13:1–327.

——. 1891. "Descriptions of New Species of Shells from the Challenger Expedition." *Proceedings of the Zoological Society of London*, 436–45.

Spry, W. J. 1878. *The Cruise of HMS Challenger*. Detroit: Craig and Taylor.

Swire, Herbert. 1938. *The Voyage of the Challenger*. London: Golden Cockerel Press.

Thomson, Wyville. 1861. "On a New Paleozoic Group of Echinodermata." *Edinburgh New Philosophical Journal* 13:106–17.

——. 1864. "Sea Lilies." *The Intellectual Observer* 6:1–11.

——. 1867. "On the 'Glass-Rope' *Hyalonema*." *The Intellectual Observer* 11:81–94.

——. 1868. "On the 'Vitreous' Sponges." *Journal of Natural History* 1:114–32.

——. 1869. "On the Depths of the Sea." *Annals and Magazine of Natural History* 4:112–24.

——. 1870a. "On Deep-Sea Climates." *Nature* 2:257–61.

——. 1870b. "On *Holtenia*, a Genus of Vitreous Sponges." *Proceedings of the Royal Society of London* 114:32–35.

——. 1871a. "The Continuity of the Chalk." *Nature* 3:225–27.

——. 1871b. "On the Distribution of Temperature in the North Atlantic." *Nature* 4:251–53.

——. 1871c. "The Relations between Zoology and Palaeontology." *Nature* 5:34–5.

——. 1871–1872a. "On the Crinoids of the 'Porcupine' Deep-Sea Dredging Expedition." *Proceedings of the Royal Society of Edinburgh* 7:764–73.

——. 1871–1872b. "On the Structure of the Palaeozoic Crinoids." *Proceedings of the Royal Society of Edinburgh* 7:415–18.

——. 1873a. *The Depths of the Sea*. London.

——. 1873b. "Notes from the 'Challenger'" [I–VII], *Nature* (May 8; May 15; June 5; July 24; July 31; August 28; September18).
——. 1874a. "The 'Challenger' in the South Atlantic." *Nature* 10:142–44.
——. 1874b. "The 'Challenger' Expedition" (I–II). *Nature* 11:95–97, 116–19.
——. 1874c. "On Dredgings and Deep-Sea Sounding in the South Atlantic." *Proceedings of the Royal Society* 22:423–28.
——. 1874d. "Preliminary Notes on the Nature of the Sea-Bottom Procured by the Soundings of H.M.S. 'Challenger' during Her Cruise in the Southern Sea." *Proceedings of the Royal Society* 22:32–49.
——. 1875a. "Dr. R. Von Willemoes-Suhm." *Nature* 13:88–9.
——. 1875b. "Report to the Hydrographer of the Admiralty on the Cruise of H.M.S. 'Challenger' from July to November 1874." *Proceedings of the Royal Society* 23:245–50.
——. 1876a. "The Challenger Expedition." *Nature* 14:492–95.
——. 1876b. "Notice of New Living Crinoids Belonging to the Apiocrinidae." *Linnean Society Journal* 13:47–55.
——. 1876c. "Preliminary Report to the Hydrographer of the Admiralty on Some of the Results of the Cruise of H.M.S. 'Challenger' between Hawaii and Valparaiso." *Proceedings of the Royal Society* 24:463–70.
——. 1876d. "Report to the Hydrographer of the Admiralty on the Cruise of H.M.S. 'Challenger' from June to August 1875." *Proceedings of the Royal Society* 24:33–40.
——. 1876e. "Report to the Hydrographer of the Admiralty on the Voyage of the 'Challenger' from the Falkland Islands to Montevideo." *Proceedings of the Royal Society* 24:623–36.
——. 1877. *The Voyage of the* Challenger: *The Atlantic*. 2 vols. London.
——. 1878. "Address." *Report of the Meeting of the British Association for the Advancement of Science*, 613–22.
——. 1880a. "General Introduction to the Zoological Series of Reports." *Scientific Results: Zoology* 1:1–59.
——. 1880b. "Geological Changes of Level." *Nature* 23:33.
Thomson, Wyville, and John Murray. 1885. *Scientific Results. Narrative*. 2 vols.
Verne, Jules. 1876. *Twenty Thousand Leagues under the Sea*. London: Samson Low.
Wild, John James. 1877. *Thalassa: An Essay on the Depth, Temperature and Currents of the Ocean*. London: Marcus Ward & Co.
——. 1878. *At Anchor: A Narrative of Experiences Afloat and Ashore during the Voyage of H.M.S. Challenger from 1872 to 1876*. London: Marcus Ward & Co.
Willemoes-Suhm, Rudolf. 1873. "On a New Genus of Amphipod Crustaceans." *Philosophical Transactions of the Royal Society of London* 163:629–38.
——. 1873–1876. "Von der Challenger-Expedition: Briefe an C.Th.E. v. Siebold von R. v. Willemoes-Suhm, I–VII." *Zeitschrift für Wissenschaftliche Zoologie* 24 [in seven parts].
——. 1876a. "On the Development of *Lepas fascicularis* and the *Archizoea* of Cirripedia." *Proceedings of the Royal Society* 24:129–32.
——. 1876b. "Preliminary Remarks on the Development of Some Pelagic Decapods." *Proceedings of the Royal Society* 24:132–34.

——. 1876c. "Preliminary Report . . . on Crustacea Observed during the Cruise of H.M.S. 'Challenger' in the Southern Sea." *Proceedings of the Royal Society* 24:585–92.

——. 1876d. "Preliminary Report . . . on Observations Made during the Earlier Part of the Voyage of H.M.S. 'Challenger.'" *Proceedings of the Royal Society* 24:569–85.

——. 1877. *Challenger-Briefe von Rudolf v. Willemoes-Suhm, 1872–1875, nach dem tode des verfassers herausgegeben von seiner mutter*. Leipzig.

SECONDARY SOURCES

Aitken, Frédéric, and Jean-Nouma Foule. 2019. *From Deep Sea to Laboratory*. 2 vols. London: Wiley.

Alaniz, John Rodolfo. 2014. "Dredging Evolutionary Theory: The Emergence of the Deep Sea as a Transatlantic Site for Evolution, 1853–1876." PhD diss., University of California, San Diego.

Bell, Graham. 2022. *Full Fathom 5000: The Expedition of HMS Challenger and the Strange Animals It Found in the Deep Sea*. Oxford: Oxford University Press.

Brunton, Eileen. 1994. *The Challenger Expedition, 1872–76: A Visual Index*. London: Natural History Museum.

Burstyn, Harold L. 1971–1972. "Pioneering in Large-Scale Scientific Organisation: The *Challenger* Expedition and Its Report. I. Launching the Expedition." *Proceedings of the Royal Society of Edinburgh* 72.3:47–61.

Charnock, H. 1973. "H.M.S. *Challenger* and the Development of Marine Science." *Journal of Navigation* 26.1:1–12.

Corfield, Richard. 2004. *The Silent Landscape: In the Wake of HMS Challenger, 1872–1876*. London: John Murray.

Deacon, Margaret. 1971. *Scientists and the Sea, 1650–1900: A Study of Marine Science*. London: Academic Press.

Deacon, Margaret, Tony Rice, and Colin Summerhayes, eds. 2001. *Understanding the Oceans: A Century of Ocean Exploration*. London: University College London Press.

Gould, W. John. 2022. "HMS *Challenger* and SMS *Gazelle*—Their 19th-century Voyages Compared." *History of the Geo- and Space Sciences* 13:171–204.

Hedgpeth, Joel W. 1946. "The Voyage of the Challenger." *Scientific Monthly* 63.3:194–202.

Herdman, William. 1923. *Founders of Oceanography and Their Work: An Introduction to the Science of the Sea*. London: Edward Arnold & Co.

Jones, Erika. 2023. *The Challenger Expedition: Exploring the Ocean's Depths*. London: Royal Museums Greenwich.

Jones, Robert Wynn. 1994. *The* Challenger *Foraminifera*. Oxford: Oxford University Press.

Lingwood, P. F. 1981. "The Dispersal of the Collections of H.M.S. *Challenger*; an Example of the Importance of Historical Research in Tracing a Systematically Important Collection." *Archives of Natural History* 1:71–77.

Linklater, Eric. 1972. *The Voyage of the Challenger*. New York: Doubleday.

MacDougall, Doug. 2019. *Endless Novelties of Extraordinary Interest: The Voyage of H.M S. Challenger and the Birth of Modern Oceanography*. New Haven: Yale University Press.

Manten, A. A. 1972. "C. Wyville Thomson, J. Murray, and the 'Challenger' Expedition." *Earth Science Reviews* 8.2:255–66.

Merriman, Daniel. 1971–1972. "Challengers of Neptune: The 'Philosophers.'" *Proceedings of the Royal Society of Edinburgh* 72.2:15–45.

Nyhart, Lynn K. 2012. "Voyaging and the Scientific Expedition Report, 1800–1940," in *Science in Print: Essays on the History of Science and the Culture of Print*, ed. Rima D. Apple, Gregory J. Downey, Stephen L. Vaughan. Madison: University of Wisconsin Press.

Pearson, Philip. 2021. *A Challenger's Song*. London: Austin Macauley.

Rehbock, Philip F. 1983. *The Philosophical Naturalists: Themes in Early Nineteenth-Century British Biology*. Madison: University of Wisconsin Press.

Rozwadowski, Helen M. 2005. *Fathoming the Ocean: The Discovery and Exploration of the Deep Sea*. Cambridge, MA: Harvard University Press.

Schlee, Susan. 1973. *The Edge of an Unfamiliar World: A History of Oceanography*. New York: Dutton.

Yonge, Maurice. 1971–1972. "The Inception and Significance of the *Challenger* Expedition." *Proceedings of the Royal Society of Edinburgh* 72.1:1–13.

Zuroski, Emma J. 2019. "Depths of Knowledge: HMS *Challenger* and the Reconfiguration of Modern Science." PhD diss., University of Auckland.

SCIENTIFIC LITERATURE (INCLUDES HISTORICAL TEXTS AND HISTORIES OF SCIENCE)

PREFACE AND INTRODUCTION

Anderson, Thomas R., and Tony Rice. 2006. "Deserts on the Sea Floor: Edward Forbes and His Azoic Hypothesis for a Lifeless Deep Ocean." *Endeavour* 30.4:131–37.

Bolster, W. Jeffrey. 2012. *The Mortal Sea: Fishing the Atlantic in the Age of Sail*. Cambridge, MA: Harvard University Press.

Erlandson, Jon. M., et al. 2007. "The Kelp Highway Hypothesis: Marine Ecology, the Coastal Migration Theory, and the Peopling of the Americas." *Journal of Island & Coastal Archaeology* 3:277–81.

Forbes, Edward. 1843. "Report of the Mollusca and Radiata of the Aegean Sea." *Report of the British Association for the Advancement of Science*, 129–93.

Forbes, Edward, and T.A.B. Spratt. 1847. *Travels in Lycia, Milyas, and the Cibyratis*. 2 vols. London: John Van Voorst.

Huxley, Thomas Henry. 1884. "Inaugural Address." *Fisheries Exhibition Literature*. London: William Clowes and Sons.

Jackson, Jeremy B. C., et al. 2001. "Historical Overfishing and the Recent Collapse of Coastal Ecosystems." *Science* 293:629–38.

McCauley, Douglas J., et al. 2015. "Marine Defaunation: Animal Loss in the Global Ocean." *Science* 347. https://doi.org/10.1126/science.1255641.

Myers, Ransom A., and Boris Worm. 2013. "Rapid Worldwide Depletion of Predatory Fish Communities." *Nature* 423:280–83.

Ramirez-Llodra, E., et al. 2010. "Deep, Diverse, and Definitely Different: Unique Attributes of the World's Largest Ecosystem." *Biogeosciences* 7:2851–99.

Rehbock, Philip F. 1979. "The Early Dredgers: 'Naturalizing' in British Seas, 1830–50." *Journal of the History of Biology* 12.2:293–368.

CHAPTER 1. EMBRITTLED STAR

Bergmann, M., et al., eds. 2015. *Marine Anthropogenic Litter*. New York: Springer.

Courtene-Jones, Winnie, et al. 2017. "Microplastic Pollution Identified in Deep-Sea Water and Ingested by Benthic Invertebrates in the Rockall Trough, North Atlantic Ocean." *Environmental Pollution* 231:271–80.

———. 2019. "Consistent Microplastic Ingestion by Deep-Sea Invertebrates over the Last Four Decades (1976–2015), a Study from the North East Atlantic." *Environmental Pollution* 244:503–12.

Cózar, Andrés, et al. 2014. "Plastic Debris in the Open Ocean." *PNAS* 111.28:10239–44.

De Pourtalès, Louis François. 1869. "Preliminary Report on the Ophiuridae and Astrophytidae Dredged in Deep Water between Cuba and the Florida Reef." *Bulletin of the Museum of Comparative Zoology* 1:309–54.

Eriksen, Marcus, et al. 2014. "Plastic Pollution in the World's Oceans: More than 5 Trillion Plastic Pieces Weighing over 250,000 Tons Afloat at Sea." *PLOS ONE* 9.12:e111913.

Ewin, Timothy A. M., and Ben Thuy. 2017. "Brittle Stars from the British Oxford Clay: Unexpected Ophiuroid Diversity on Jurassic Sublittoral Mud Bottoms." *Journal of Paleontology* 91.4:781–98.

Fišer, Cene, et al. 2017. "Cryptic Species as a Window into the Paradigm Shift of the Species Concept." *Molecular Ecology* 27:613–35.

Forbes, Edward. 1841. *A History of British Starfishes*. London: John Van Voorst.

———. 1844. "On the Fossil Remains of Starfishes of the Order Ophiuridae, Found in Britain." *Proceedings of the Geological Society of London* 4:232–34.

———. 1859. *A History of the European Seas*. London: John Van Voorst.

Gage, J. D., and P. A. Tyler. 1982. "Growth and Reproduction of the Deep-sea Brittlestar *Ophiomusium lymani* Wyville Thomson." *Oceanologica Acta* 5.1:73–83.

Goss, Richard J. 1992. "The Evolution of Regeneration: Adaptive or Inherent?" *Journal of Theoretical Biology* 159:241–60.

Gregory, Murray R. 2009. "Environmental Implications of Plastic Debris in Marine Settings—Entanglement, Ingestion, Smothering, Hangers-On, Hitch-Hiking and Alien Invasions." *Philosophical Transactions of the Royal Society B* 364:2013–25.

Koelmans, Albert A., et al. 2017. "All Is Not Lost: Deriving a Top-Down Budget of Plastic at Sea." *Environmental Research Letters* 12:114028.

Kooi, Merel, et al. 2017. "Ups and Downs in the Ocean: Effects of Biofouling on Vertical Transport of Microplastics." *Environmental Science and Technology* 51:7963–71.

Mills, Eric. 1978. "Edward Forbes, John Gwyn Jeffreys, and British Dredging before the *Challenger* Expedition." *Journal of the Society for the Bibliography of Natural History* 8.4:507–36.

———. 1984. "A View of Edward Forbes, Naturalist." *Archives of Natural History* 11.3:365–93.

O'Hara, Timothy D., et al. 2014. "Phylogenomic Resolution of the Class Ophiuroidea Unlocks a Global Microfossil Record." *Current Biology* 24:1874–79.

———. 2019. "Contrasting Processes Drive Ophiuroid Phylodiversity across Shallow and Deep Seafloors." *Nature* 565:636–39.

Pearson, M., and J. D. Gage. 1984. "Diets of Some Deep-Sea Brittle Stars in the Rockall Trough." *Marine Biology* 82:247–58.

Stöhr, Sabine, et al. 2012. "Global Diversity of Brittle Stars (Echinodermata: Ophiuroidea)." *PLOS ONE* 7.3:e31940.

Teuten, Emma L., et al. 2009. "Transport and Release of Chemicals from Plastics to the Environment and to Wildlife." *Philosophical Transactions of the Royal Society B* 364:2027–45.

Thuy, Ben, et al. 2012. "Ancient Origin of the Modern Deep-Sea Fauna." *PLOS ONE* 7.10: e46913.

———. 2014. "First Glimpse into Lower Jurassic Deep-Sea Diversity: *In Situ* Diversification and Resilience against Extinction." *Proceedings of the Royal Society B* 281:20132624.

———. 2021. "New Fossils of Jurassic Ophiurid Brittle Stars (Ophiuroidea; Ophiurida) Provide Evidence for Early Clade Evolution in the Deep Sea." *Royal Society Open Science* 8:210643.

Tyler, P. A., and J. D. Gage. 1979. "Reproductive Ecology of Deep Sea Ophiuroids from the Rockall Trough," in *Cyclic Phenomena in Marine Plants and Animals*, ed. E. Naylor and R. G. Hartness. Oxford: Pergamon Press, 215–22.

Van Cauwenberghe, et al. 2013. "Microplastic Pollution in Deep-Sea Sediments." *Environmental Pollution* 182:495–99.

Wilkie, I. C. 2001. "Autotomy as a Prelude to Regeneration in Echinoderms." *Microscopy Research and Technique* 55:369–96.

Woodall, Lucy C., et al. 2014. "The Deep Sea Is a Major Sink for Microplastic Debris." *Royal Society Open Science* 1:140317.

Woolley, Skipton N. C., et al. 2016. "Deep-Sea Diversity Patterns Are Shaped by Energy Availability." *Nature* 533:393–96.

CHAPTER 2. SKELETONS FROM THE OOZE

Anderson, David M., et al. 2002. "Increase in the Asian Southwest Monsoon during the Past Four Centuries." *Science* 297:596–99.

Beaugrand, Gregory, et al. 2013. "Long-term Responses of North Atlantic Calcifying Plankton to Climate Change." *Nature Climate Change* 3:263–67.

Di Gregorio, Mario A. 2005. *From Here to Eternity: Ernst Haeckel and Scientific Faith*. Göttingen: Vandenhoeck & Ruprecht.

Ehrenberg, Christian. 1874. "Grössere Felsproben des Polycystinen-Mergels von Barbados mit weiteren Erläuterungen." *Monatsberichte der Königlichen Preussische Akademie des Wissenschaften zu Berlin*, 213–62.

Fox, Lyndsey, Stephen Stukins, Thomas Hill, and C. Giles Miller. 2020. "Quantifying the Effect of Anthropogenic Climate Change on Calcifying Plankton." *Nature Scientific Reports* 10:1620.

Gupta, Anil K., et al. 2015. "Evolution of the South Asian Monsoon Wind System since the Late Middle Miocene." *Palaeogeography, Palaeoclimatology, Palaeoecology* 438:160–67.

Jonkers, Lucas, et al. 2019. "Global Change Drives Modern Plankton Communities Away from the Pre-industrial State." *Nature* 570:372–75.

Moy, Andrew D., et al. 2009. "Reduced Calcification in Modern Southern Ocean Planktonic Foraminifera." *Nature Geoscience* 2:276–80.

Ohtsuka, M., et al., eds. 2015. *Marine Protists*. New York: Springer Japan.

Richards, Robert. 2008. *The Tragic Sense of Life: Ernst Haeckel and the Struggle over Evolutionary Thought*. Chicago: University of Chicago Press.

Sen Gupta, Barun K., ed. 1999. *Modern Foraminifera*. London: Kluwer Academic Publishers.

Shrivastav, Ankush, et al. 2016. "Significance of *Globigerina bulloides* D'Orbigny: A Foraminiferal Proxy for Palaeomonsoon and Past Upwelling Records." *Journal of Climate Change* 2.2:99–110.

CHAPTER 3. THE FIRE REEFS

Albright, Rebecca, et al. 2010. "Ocean Acidification Compromises Recruitment Success of the Threatened Caribbean Coral *Acropora palmata*." *PNAS* 107.47:20400–404.

Alvarez-Filip, Lorenzo, et al. 2009. "Flattening of Caribbean Coral Reefs: Region-Wide Declines in Architectural Complexity." *Proceedings of the Royal Society B* 276:3019–25.

Aronson, Richard B., and William F. Precht. 2001. "White-Band Disease and the Changing Face of Caribbean Coral Reefs." *Hydrobiologia* 460:25–38.

Cook, Clayton, et al. 1990. "Elevated Temperatures and Bleaching on a High Latitude Coral Reef: The 1988 Bermuda Event." *Coral Reefs* 9:45–49.

Droxler, André, and Stéphan J. Jorry. 2021. "The Origin of Modern Atolls: Challenging Darwin's Deeply Ingrained Theory." *Annual Review of Marine Science* 13:537–73.

Fitt, William K. 2012. "Bleaching of the Fire Corals *Millepora*." *Proceedings of the 12th International Coral Reef Symposium*, 9A.

Hewson, Ian. 2023. "A Scuticociliate Causes Mass Mortality of *Diadema antillarum* in the Caribbean Sea." *Science Advances* 9:eadg3200.

Jackson, J.B.C. 1997. "Reefs Since Columbus." *Coral Reefs* 16:23–32.

Lessios, H. A. 2016. "The Great *Diadema antillarum* Die-Off: 30 Years Later." *Annual Review of Marine Science* 8:267–83.

Lessios, H. A., et al. 1984. "Spread of *Diadema* Mass Mortality through the Caribbean." *Science* 226:335–37.

Lewis, John B. 2006. "Biology and Ecology of the Hydrocoral *Millepora* on Coral Reefs." *Advances in Marine Biology* 50:1–55.

Lundgren, Ian, and Zandy Hillis-Starr. 2008. "Variation in *Acropora palmata* Bleaching across Benthic Zones at Buck Island National Monument (St. Croix, USVI) during the 2005 Thermal Stress Event." *Bulletin of Marine Science* 83.3:441–51.

Miller, M. W., et al. 2014. "Prevalence, Consequences, and Mitigation of Fireworm Predation on Endangered Staghorn Coral." *Marine Ecology Progress Series* 516:187–94.

Muller, Erin M., et al. 2008. "Bleaching Increases Likelihood of Disease on *Acropora palmata* (Lamarck) in Hawksnest Bay, St. John, US Virgin Islands." *Coral Reefs* 27:191–95.

———. 2018. "Bleaching Causes Loss of Disease Resistance within the Threatened Coral Species *Acropora cervicornis*." *eLife* 7:e35066.

Paddack, Michelle J., et al. 2009. "Recent Region-wide Declines in Caribbean Reef Fish Abundance." *Current Biology* 19:590–95.

Rice, Mallory M., et al. 2019. "Corallivory in the Anthropocene: Interactive Effects of Anthropogenic Stressors and Corallivory on Coral Reefs." *Frontiers in Marine Science* 5:A525.

Ritchie, Kim B. 2006. "Regulation of Microbial Populations by Coral Surface Mucus and Mucus-Associated Bacteria." *Marine Ecology Progress Series* 322:1–14.

Rogers, C. S., and E. M. Muller. 2012. "Bleaching, Disease, and Recovery in the Threatened Scleractinian Coral *Acropora palmata* in St. John, US Virgin Islands: 2003–2010." *Coral Reefs* 31:807–19.

Schulze, Anja, et al. 2017. "Tough, Armed, and Omnivorous: *Hermodice carunculata* (Annelida: Amphinomidae) Is Prepared for Ecological Challenges." *Journal of the Marine Biological Association of the United Kingdom* 97.5:1075–80.

Sussman, Meir, et al. 2003. "The Marine Fireworm *Hermodice carunculata* Is a Winter Reservoir and Spring-Summer Vector for the Coral-Bleaching Pathogen *Vibrio shiloi*." *Environmental Microbiology* 5.4:250–55.

Sutherland, Kathryn P., et al. 2011. "Human Pathogen Shown to Cause Disease in the Threatened Elkhorn Coral *Acropora palmata*." *PLOS ONE* 6.8:e23468.

Sutherland, Kathryn P., et al. 2016. "Shifting White Pox Aetiologies Affecting *Acropora palmata* in the Florida Keys, 1994–2014." *Philosophical Transactions of the Royal Society B* 371.

Williams, D. E., et al. 2017. "Thermal Stress Exposure, Bleaching Response, and Mortality in the Threatened Coral *Acropora palmata*." *Marine Pollution Bulletin* 24:189–97.

CHAPTER 4. WIDE SARGASSO SEAWEED

Devault, Damien A., et al. 2021. "The Silent Spring of *Sargassum*." *Environmental Science and Pollution Research* 28:15580–83.

García-Sánchez, Marta, et al. 2020. "Temporal Changes in the Composition and Biomass of Beached Pelagic *Sargassum* Species in the Mexican Caribbean." *Aquatic Botany* 167:103275.

Johns, Elizabeth M., et al. 2020. "The Establishment of a Pelagic *Sargassum* Population in the Tropical Atlantic: Biological Consequences of a Basin-Scale Long Distance Dispersal." *Progress in Oceanography* 188:102437.

Laffoley, Dan, et al. 2011. *The Protection and Management of the Sargasso Sea.* Sargasso Sea Alliance.

Oviatt, Candace A., et al. 2019. "What Nutrient Sources Support Anomalous Growth and the Recent Sargassum Mass Stranding on Caribbean Beaches? A Review." *Marine Pollution Bulletin* 145:517–25.

Rodríguez-Martínez, R. E., et al. 2019. "Faunal Mortality Associated with Massive Beaching and Decomposition of Pelagic *Sargassum*." *Marine Pollution Bulletin* 146:201–5.

Smetacek, Victor, and Adriana Zingone. 2013. "Green and Golden Seaweed Tides on the Rise." *Nature* 504:84–88.

Wang, Mengqiu, et al. 2019. "The Great Atlantic *Sargassum* Belt." *Science* 365:83–87.

CHAPTER 5. DARKNESS VISIBLE

Bagge, Laura, et al. 2016. "Nanostructures and Monolayers of Spheres Reduce Surface Reflections in Hyperiid Amphipods." *Current Biology* 26:3071–76.

Baum, J. K., and A.C.J. Vincent. 2005. "Magnitude and Inferred Impacts of the Seahorse Trade in Latin America." *Environmental Conservation* 32:305–19.

Baum, J. K., et al. 2003. "Bycatch of Lined Seahorses (*Hippocampus erectus*) in a Gulf of Mexico Shrimp Trawl Fishery." *Fishery Bulletin* 101:721–31.

Chen, Lu, et al. 2015. "The Genus *Hippocampus*—A review of Traditional Medicinal Uses, Chemical Constituents and Pharmacological Properties." *Journal of Ethnopharmacology* 162:104–11.

Cronin, Thomas W. 2016. "Camouflage: Being Invisible in the Open Ocean." *Current Biology* 26:R1179–R1181.

Davies, Thomas W., et al. 2014. "The Nature, Extent, and Ecological Implications of Marine Light Pollution." *Frontiers in Ecology and the Environment* 12.6:347–55.

Duarte, Michele, et al. 2019. "Disruptive Coloration and Habitat Use by Seahorses." *Neotropical Icthyology* 17.4:e190064.

Duarte, Rafael, et al. 2017. "Camouflage through Colour Change: Mechanisms, Adaptive Value and Ecological Significance." *Philosophical Transactions of the Royal Society B* 372:20160342.

Foster, S. J., and A.C.J. Vincent. 2004. "Life History and Ecology of Seahorses: Implications for Conservation and Management." *Journal of Fish Biology* 65:1–61.

Gosse, Philip Henry. 1853. *A Naturalist's Rambles on the Devonshire Coast*. London: John Van Voorst.

———. 1854. *The Aquarium: An Unveiling of the Wonders of the Deep Sea*. London: John Van Voorst.

Haddock, Steven H. D., et al. 2010. "Bioluminescence in the Sea." *Annual Review of Marine Science* 2:443–93.

Johnsen, Sonke. 2001. "Hidden in Plain Sight: The Ecology and Physiology of Organismal Transparency." *Biology Bulletin* 201:301–18.

———. 2003. "Lifting the Cloak of Invisibility: The Effects of Changing Optical Conditions on Pelagic Crypsis." *Integrated Comparative Biology* 43:580–90.

———. 2005. "The Red and the Black: Bioluminescence and the Color of Animals in the Deep Sea." *Integrated Comparative Biology* 45:234–46.

Johnsen, Sonke, et al. 2004. "Propagation and Perception of Bioluminescence: Factors Affecting Counterillumination as a Cryptic Strategy." *Biology Bulletin* 207:1–16.

Lampert, W. 1989. "The Adaptive Significance of Diel Vertical Migration of Zooplankton." *Functional Biology* 3:21–27.
Meester, L. D. 2009. "Diel Vertical Migration," in *Encyclopedia of Inland Waters*, ed. Gene E. Likens. Amsterdam: Elsevier.
Piontkovski, S. A. 1997. "The Bioluminescent Field of the Atlantic Ocean." *Marine Ecology Progress Series* 156:33–41.
Rosa, Ierecê L., et al. "Collaborative Monitoring of the Ornamental Trade of Seahorses and Pipefishes (Teleostei: Syngnathidae) in Brazil: Bahia State as a Case Study." *Neotropical Icthyology* 4.2:247–52.
Scales, Helen. 2009. *Poseidon's Steed: The Story of Seahorses, from Myth to Reality*. New York: Gotham Books.
——. 2010. "Advances in the Ecology, Biogeography and Conservation of Seahorses (Genus *Hippocampus*)." *Progress in Physical Geography* 34.4:443–58.
"The Sea-Horse (*Hippocampus brevirostris*)." 1874. *Hardwicke's Science Gossip*, ed. J. E. Taylor. London: Hardwicke.
Vincent, A.J.C., et al. 2011. "Conservation and Management of Seahorses and Other Syngnathidae." *Journal of Fish Biology* 78:1681–1724.
Warren, Eric J., and N. Adam Locket. 2004. "Vision in the Deep Sea." *Biological Reviews* 79:671–712.
Widder, E. A. 2010. "Bioluminescence in the Ocean: Origins of Biological, Chemical, and Ecological Diversity." *Science* 328:704–8.
——. 2021. *Below the Edge of Darkness: A Memoir of Exploring Light and Life in the Deep Sea*. New York: Random House.
Wilson, Thérèse, and J. Woodland Hastings. 2013. *Bioluminescence: Living Lights, Lights for Living*. Cambridge, MA: Harvard University Press.

CHAPTER 6. HEYDAY OF THE OCTOPUS

Anderson, Sean C., et al. 2011. "Rapid Global Expansion of Invertebrate Fisheries: Trends, Drivers, and Ecosystem Effects." *PLOS ONE* 6.3:e14735.
André, Jessica, et al. 2009. "Effects of Temperature on Energetics and the Growth Pattern of Benthic Octopuses." *Marine Ecology Progress Series* 374:167–79.
Aristotle. 1984. "Historia Animalium," in *Complete Works*, ed. Jonathan Barnes. Princeton: Princeton University Press, 1:774–993.
Brotz, Lucas, et al. 2012. "Increasing Jellyfish Populations: Trends in Large Marine Ecosystems." *Hydrobiologia* 690:3–20.
Collins, Martin A., and Paul G. K. Rodhouse. 2006. "Southern Ocean Cephalopods." *Advances in Marine Biology* 50:191–265.
Doubleday, Zoë, et al. 2016. "Global Proliferation of Cephalopods." *Current Biology* 26:406–7.
Doubleday, Zoë, and Sean D. Connell. 2018. "Weedy Futures: Can We Benefit from the Species That Thrive in the Marine Anthropocene?" *Frontiers in Ecology and the Environment* 16.10:599–604.
Duarte, Carlos M., et al. 2013. "Is Global Ocean Sprawl a Cause of Jellyfish Blooms?" *Frontiers in Ecology and the Environment* 11.2:91–97.

Gould, W. John, and Stuart A. Cunningham. 2021. "Global-Scale Patterns of Observed Sea Surface Salinity Intensified since the 1870s." *Communications Earth and Environment* 2:76.

Onthank, Kirt, et al. 2021. "Impact of Short and Long-Term Exposure to Elevated Pco2 on Metabolic Rate and Hypoxia Tolerance in *Octopus rubescens*." *Physiological and Biochemical Zoology* 94.1:1–11.

Parsons, T., and C. M. Lalli. 2002. "Jellyfish Population Explosions: Revisiting a Hypothesis of Possible Causes." *La Mer* 40:111–21.

Purcell, Jennifer E. 2012. "Jellyfish and Ctenophore Blooms Coincide with Human Proliferations and Environmental Perturbations." *Annual Review of Marine Science* 4:209–35.

Richardson, Anthony J., et al. 2009. "The Jellyfish Joyride: Causes, Consequences and Management Responses to a More Gelatinous Future." *Trends in Ecology and Evolution* 24.6:312–22.

Rodhouse, Paul G. K. 2013. "Role of Squid in the Southern Ocean Pelagic System and the Possible Consequences of Climate Change." *Deep Sea Research II* 95:129–38.

Rodhouse, Paul G. K., and J. F. Caddy. 1998. "Cephalopod and Groundfish Landings: Evidence for Ecological Change in Global Fisheries?" *Reviews in Fish Biology and Fisheries* 8:431–44.

Wray, Phoebe, and Kenneth R. Martin. 1980. "Historical Whaling Records from the Western Indian Ocean." *Report of the International Whaling Commission* 5:213–31.

CHAPTER 7. BROOCH CLAMS AND HAIRY MUSSELS

Agassiz, Louis. 1840–1845. *Etudes Critiques sur les Mollusques Fossiles*. 3 vols. Neuchâtel: Imprimerie de Petitpiere.

Alleway, Heidi K., and Sean D. Connell. 2015. "Loss of an Ecological Baseline through the Eradication of Oyster Reefs from Coastal Ecosystems and Human Memory." *Conservation Biology* 29.3:795–804.

Beesley, P. L., et al., eds. 1998. *Mollusca: The Southern Synthesis*. 2 vols. Melbourne: CSIRO Publishing.

Dumont D'Urville, Jules-Sébastien-César. 1830–1834. *Voyage de la corvette l'Astrolabe . . . pendant les années 1826, 1827, 1828, 1829*. 5 vols. Paris: J. Tatsu.

Ford, John R., and Paul Hamer. 2016. "The Forgotten Shellfish Reefs of Coastal Victoria: Documenting the Loss of a Marine Ecosystem over 200 Years since European Settlement." *Royal Society of Victoria* 128:87–105.

Gillies, Chris L., et al. 2018. "Australian Shellfish Ecosystems: Past Distribution, Current Status and Future Direction." *PLOS ONE* 13.2:e0190914.

Gould, Stephen Jay. 1968. "*Trigonia* and the Origin of Species." *Journal of the History of Biology* 1:41–56.

Jenkins, H. M. 1865. "On the Occurrence of a Tertiary Species of Trigonia in Australia." *Quarterly Journal of Science* 2:362–64.

Johnson, Craig R., et al. 2011. "Climate Change Cascades: Shifts in Oceanography, Species' Ranges and Subtidal Marine Community Dynamics in Eastern Tasmania." *Experimental Marine Biology and Ecology* 400:17–32.

Kirby, Michael X. 2004. "Fishing Down the Coast: Historical Expansion and Collapse of Oyster Fisheries along Continental Margins." *PNAS* 101.35:13096–99.

Lamarck, Jean Baptiste. 1799. "Prodrome d'une Nouvelle Classification des Coquilles." *Mémoires de la Societé Histoire Naturelle de Paris* 1:63–91.

———. 1804. "Sur une Nouvelle Espèce de Trigonie. . . ." *Annales du Muséum d'histoire naturelle* 4:351–59.

Ling, S. D. 2008. "Range Expansion of a Habitat-Modifying Species Leads to Loss of Taxonomic Diversity: A New and Impoverished Reef State." *Oecologica* 156.4:883–94.

———. 2009. "Climate-Driven Range Extension of a Sea Urchin: Inferring Future Trends by Analysis of Recent Population Dynamics." *Global Change Biology* 15.3:719–31.

Lubbock, John. 1869. *Pre-historic Times*. 2nd ed. London: Williams & Norgate.

Meehan, Betty. 1982. *Shell Bed to Shell Midden*. Canberra: Australian Institute of Aboriginal Studies.

Ogburn, Damian, et al. 2007. "The Disappearance of Oyster Reefs from Eastern Australian Estuaries—Impact of Colonial Settlement or Mudworm Invasion?" *Coastal Management* 35:217–87.

Oliver, E.C.J., et al. 2014. "Projected Tasman Sea Extremes in Sea Surface Temperature through the Twenty-First Century." *Journal of Climate* 27.5:1980–98.

Péron, François. 1807. *Voyages de découvertes aux terres Australes . . . pendant les années 1800, 1801, 1802, 1803 et 1804*. 2 vols. Paris: De L'Imprimerie Impériale.

Ridgeway, K. R., and J. R. Dunn. 2007. "Observational Evidence for a Southern Hemisphere Oceanic Supergyre." *Geophysical Research Letters* 34.13:L13612.

Roemmich, D., et al. 2007. "Decadal Spinup of the South Pacific Tropical Gyre." *Journal of Physical Oceanography* 37.2:112–23.

Soon, Tan Kar, and Juaiping Zheng. 2019. "Climate Change and Bivalve Mass Mortality in Temperate Regions." *Reviews of Environmental Contamination and Toxicology* 251:109–29.

Thomson, Lindsay G., ed. 1893. *History of the Fisheries of New South Wales*. Sydney: C. Potter.

Wahl, M., et al. 2018. "Macroalgae May Mitigate Ocean Acidification Effects on Mussel Calcification by Increasing pH and Its Fluctuations." *Limnology and Oceanography* 63.1:3–21.

Wilson, Laura J., et al. 2016. "Climate-Driven Changes to Ocean Circulation and Their Inferred Impacts on Marine Dispersal Patterns." *Global Ecology and Biogeography* 25:923–39.

CHAPTER 8. THE RECKLESSNESS OF BEAUTY

Aizenberg, Joanna, et al. 2005. "Skeleton of *Euplectella* sp.: Structural Hierarchy from the Nanoscale to the Macroscale." *Science* 309:275–78.

Chimmo, William. 1878. *On* Euplectella Aspergillum. London: Taylor & Francis.

Dance, S. Peter. 1980. "Hugh Cuming, Prince of Collectors." *Journal of the Society for the Bibliography of Natural History* 9.4:477–501.

Desmond, Adrian. 1984. "Robert E. Grant: The Social Predicament of a Pre-Darwinian Transmutationist." *Journal of the History of Biology* 17.2:189–223.

Fernandes, Matheus C. 2021. "Mechanically Robust Lattices Inspired by Deep-Sea Glass Sponges." *Nature Materials* 20:237–41.

Grant, Robert. 1825–1827. "Observations and Experiments on the Structure and Functions of the Sponge." *Edinburgh Philosophical Journal* 13:94–107; 13:333–46; 14:113–24; *Edinburgh New Philosophical Journal* 2:121–41.

———. 1826. "On the Structure and Nature of the *Spongilla friabilis*." *Edinburgh Philosophical Journal* 14:270–84.

———. 1826. "Observations on the Structure of some Silicious Sponges." *Edinburgh New Philosophical Journal* 1:341–51.

———. 1826. "Remarks on the Structure of Some Calcareous Sponges." *Edinburgh New Philosophical Journal* 1:166–70.

Leys, S. R., et al. 2007. "The Biology of Glass Sponges." *Advances in Marine Biology* 52:1–145.

Moore, Thomas J. 1869. "On the Habitat of the Regadera (Watering-pot) or Venus's Flower Basket." *Annals and Magazine of Natural History* 3:196–99.

Owen, Richard. 1843. "Description of a New Genus and Species of Sponge (*Euplectella aspergillum*)." *Transactions of the Zoological Society of London* 3:203–5. See also, *Proceedings of the Zoological Society of London* 9 (1841):3–5.

Robson Brown, K., et al. 2019. "The Structural Efficiency of the Sea Sponge *Euplectella aspergillum* Skeleton: Bio-inspiration for 3D Printed Architectures." *Journal of the Royal Society Interface* 16:20180965.

Saito, Tomomi, et al. 2002. "Skeletal Growth of the Deep-Sea Hexactinellid Sponge *Euplectella oweni*, and Host Selection by the Symbiotic Shrimp *Spongicola japonica*." *Journal of the Zoological Society of London* 258:521–29.

Secord, James. 1991. "Edinburgh Lamarckians: Robert Jameson and Robert E. Grant." *Journal of the History of Biology* 24.1:1–18.

Sundar, Vikram C., et al. 2003. "Fibre-Optical Features of a Glass Sponge." *Nature* 424:899–90.

Weaver, James C., et al. 2007. "Hierarchical Assembly of the Siliceous Skeletal Lattice of the Hexactinellid Sponge *Euplectella aspergillum*." *Journal of Structural Biology* 158:93–106.

CHAPTER 9. THE CHALLENGER DEEP

Bell, James J., et al. 2018. "Sponges to Be Winners under Near-Future Climate Scenarios." *BioScience* 68.12:955–68.

Benton, Michael J. 2009. "The Red Queen and the Court Jester: Species Diversity and the Role of Biotic and Abiotic Factors through Time." *Science* 323:728–32.

Bowler, Peter J. 1983. *The Eclipse of Darwinism*. Baltimore: Johns Hopkins University Press.

Chiba, Sanae, et al. 2018. "Human Footprint in the Abyss: 30 Year Records of Deep-Sea Plastic Debris." *Marine Policy* 96:204–12.

Darwin, Charles. 1880. "Sir Wyville Thomson and Natural Selection." *Nature* 23:32.

———. 1903. *More Letters of Charles Darwin*. 2 vols. Ed. Francis Darwin. New York: Appleton & Co.

Dasgupta, S., et al. 2018. "Toxic Anthropogenic Pollutants Reach the Deepest Ocean on Earth." *Geochemical Perspective Letters* 7:22–26.

Du, Mengran, et al. 2021. "Geology, Environment, and Life in the Deepest Part of the World's Oceans." *Innovation* 2.2:100109.

Huxley, Thomas Henry. 1900. *Life and Letters*. 2 vols. Ed. Leonard Huxley. London: Macmillan.

Jambeck, Jenna R., et al. 2015. "Plastic Waste Inputs from Land into the Ocean." *Science* 347:768–71.

Jamieson, Alan J., et al. 2017. "Bioaccumulation of Persistent Organic Pollutants in the Deepest Ocean Fauna." *Nature Ecology & Evolution* 1:0051.

Jamieson, Alan, et al. 2019. "Microplastics and Synthetic Particles Ingested by Deep-Sea Amphipods in Six of the Deepest Marine Ecosystems on Earth." *Royal Society Open Science* 6:180667.

Peng, Guyu, et al. 2020. "The Ocean's Ultimate Trashcan: Hadal Trenches as Major Depositories for Plastic Pollution." *Water Research* 168:115121.

Stuart, Carol T., et al. 2003. "Large-Scale Spatial and Temporal Patterns of Deep-Sea Benthic Species Diversity," in *Ecosystems of the Deep Oceans*, ed. P. A. Tyler. London: Elsevier, 295–311.

CHAPTER 10. A SALMON'S PILGRIMAGE

Blakiston, Thomas W. 1883. *Japan in Yezo*. Yokohama: Japan Gazette.

Hasegawa, Koh, et al. 2023. "Effects of Domestication and Captive Breeding on Reaction to Moving Objects: Implications for Avoidance Behaviours of Masu Salmon *Oncorhynchus masou*." *Royal Society Open Science* 10:230045.

Kaeriyama, Masahide. 2023. "Warming Climate Impacts on Production Dynamics of Southern Populations of Pacific Salmon in the North Pacific Ocean." *Fisheries Oceanography* 32:121–32.

Kaeriyama, Masahide, et al. 2009. "Trends in Run Size and Carrying Capacity of Pacific Salmon in the North Pacific Ocean." *North Pacific Anadromous Fish Commission Bulletin* 5:293–302.

Kaeriyama, Masahide, and Rizalita R. Edpalina. 2004. "Evaluation of the Biological Interaction between Wild and Hatchery Population for Sustainable Fisheries Management of Pacific Salmon," in *Stock Enhancement and Sea Ranching: Pitfalls and Opportunities*, 2nd ed., ed. K. M. Leber, S. Kitada, H. L. Blankenship, and T. Svåsand Sand. London: Blackwell, 245–59.

Kalland, Arne. 1995. *Fishing Villages in Tokugawa Japan*. Honolulu: University of Hawaii Press.

Kato, Fumihiko. 1991. "Life Histories of Masu and Amago Salmon," in *Pacific Salmon Life Histories*, ed. C. Groot and L. Margolis. Vancouver: University of British Columbia Press, 449–521.

Kobayashi, T. 1980. "Salmon Propagation in Japan," in *Salmon Ranching*, ed. J. E. Thorpe. New York: Academic Press, 91–107.

Morita, Kentaro, and Masa-aki Fukuwaka. 2007. "Why Age and Size at Maturity Have Changed in Pacific Salmon." *Marine Ecology Progress Series* 335:289–94.

Muscolino, Micah. 2013. "Fisheries Build Up the Nation: Maritime Environmental Encounters between Japan and China," in *Japan at Nature's Edge: The Environmental Context of a Global Power*, ed. Ian J. Miller, Julia A. Thomas, and Brett L. Walker. Honolulu: University of Hawaii Press, 56–70.

Nakae, Masanori, et al. 2022. "Domestication of Captive-Bred Masu Salmon *Oncorhynchus masou masou* (Salmonidae) Leads to a Significant Decrease in Numbers of Lateral Line Organs." *Nature: Scientific Reports* 12:16780.

Reinhardt, Ulrich G. 2001. "Selection for Surface Feeding in Farmed and Sea-Ranched Masu Salmon Juveniles." *American Fisheries Society* 130:155–58.

Terui, Akira, et al. 2023. "Intentional Release of Native Species Undermines Ecological Stability." *PNAS* 120.7:e2218044120.

Tsutsui, William M. 2013. "The Pelagic Empire: Reconsidering Japanese Expansion," in *Japan at Nature's Edge: The Environmental Context of a Global Power*, ed. Ian J. Miller, Julia A. Thomas, and Brett L. Walker. Honolulu: University of Hawaii Press, 21–38.

Urabe, Hirokazu, et al. 2014. "Application of a Bioenergetics Model to Estimate the Influence of Habitat Degradation by Check Dams and Potential Recovery of Masu Salmon Populations." *Environmental Biology of Fishes* 97:587–98.

Yamamoto, Toshiaki, and Ulrich G. Reinhardt. 2003. "Dominance and Predator Avoidance in Domesticated and Wild Masu Salmon *Oncorhynchus masou*." *Fisheries Science* 69:88–94.

Yonemoto, Marcia. 1999. "Maps and Metaphors of the 'Small Eastern Sea' in Tokugawa Japan (1603–1868)." *Geographical Review* 89.2:169–87.

Yu, Jeong-Nam, et al. 2012. "Genetic Differentiation between Collections of Hatchery and Wild Masu Salmon (*Oncorhynchus masou*) Inferred from Mitochondrial and Microsatellite DNA Analyses." *Environmental Biology of Fishes* 94:259–71.

CHAPTER 11. DEATH OF A NATURALIST

Adler, Antony, and Erik Dücker. 2018. "When Pasteurian Science Went to Sea: The Birth of Marine Microbiology." *Journal of the History of Biology* 51:107–33.

Anderson, D. T. 1994. *Barnacles: Structure, Function, Development, and Evolution*. London: Chapman & Hall.

Barazandeh, Marjan, et al. 2013. "Something Darwin Didn't Know about Barnacles: Spermcast Mating in a Common Stalked Species." *Proceedings of the Royal Society B* 280:20122919.

Darwin, Charles. 1851. *A Monograph on the Sub-Class* Cirripedia*: The* Lepadidae *or Pedunculated Cirripedes*. London: Ray Society.

———. 1873. "On the Males and Complemental Males of Certain Cirripedes, and on Rudimentary Structures." *Nature* 8:431–32.

Dekov, Vesselin M., et al. 2010. "Metalliferous Sediments from the HMS *Challenger* Voyage (1872–1876)." *Geochimica et Cosmochimica Acta* 74:5019–38.

Göllnitz, Martin. 2017. "Weltumsegelung mit Karriereblick: Die britische Challenger-Expedition und Rudolf von Willemoes-Suhm," in *Mit Forscherdrang und Abenteuerlust: Expeditions-und Forschungsreisen kieler Wissenschaftler und Wissenschaftlerinnen im 19. und 20. Jahrhundert*, ed. Oliver Auge and Martin Göllnitz. Frankfurt: Peter Lang, 37–65.

Haeckel, Ernst. 1869. "Monographie der Moneren," trans. W. Kirby and E. Wright. *Quarterly Journal of Microscopical Science* 9:27–42, 113–34, 219–32, 327–42.

Huxley, Thomas Henry. 1868. "On Some Organisms Living at Great Depths in the North Atlantic Ocean." *Quarterly Journal of Microscopical Science* 8:203–12.

———. 1875. "Notes from the 'Challenger.'" *Nature* 12:315–16.

Karner, Markus B., et al. 2001. "Archeal Dominance in the Mesopelagic Zone of the Pacific Ocean." *Nature* 409:507–10.

Kutschera, U. 2016. "Haeckel's 1866 Tree of Life and the Origin of Eukaryotes." *Nature Microbiology* 1:16114.

Martin, William, et al. 2008. "Hydrothermal Vents and the Origin of Life." *Nature Reviews: Microbiology* 6:805–14.

Moyer, Craig L., et al. 1995. "Phylogenetic Diversity of the Bacterial Community from a Microbial Mat at an Active Hydrothermal Vent System, Loihi Seamount, Hawaii." *Applied and Environmental Microbiology* 61.4:1555–62.

Newman, William A. 1993. "Darwin and Cirripedology," in *History of Carcinology*, ed. Frank Truesdale. Rotterdam: A. A. Balkema, 349–434.

Rehbock, Philip F. 1975. "Huxley, Haeckel, and the Oceanographers: The Case of *Bathybius haeckelii*." *Isis* 66.4:504–33.

Srinivasan, U. Thara, et al. 2012. "Global Fisheries Losses at the Exclusive Economic Zone Level, 1850 to Present." *Marine Policy* 36:544–49.

Stott, Rebecca. 2003. *Darwin and the Barnacle*. New York: W. W. Norton.

Thompson, J. V. 1830. "On the Cirripedes or Barnacles: Demonstrating Their Deceptive Character; the Extraordinary Metamorphoses They Undergo; and the Class of Animals to which They Indisputably Belong," in *Zoological Researches and Illustrations*. Cork: King and Ridings, 69–82.

Whitehead, T. Otto, et al. 2011. "South African Pelagic Goose Barnacles (Cirripedia: Thoracica): Substratum Preferences and Influence of Plastic Debris on Abundance and Distribution." *Crustaceana* 84.5–6:635–49.

Yamaguchi, Sachi, et al. 2012. "Sexual Systems and Life History of Barnacles: A Theoretical Perspective." *Integrative and Comparative Biology* 32.3:356–65.

Zheden, Vanessa, et al. 2015. "Characterization of Cement Float Buoyancy in the Stalked Barnacle *Dosima fascicularis*." *Interface Focus* 5:20140060.

CHAPTER 12. MESSAGE FROM THE COSMOS

Alarcón-Muñoz, Ruben. 2008. "Jumbo Squid (*Dosidicus gigas*) Biomass off Central Chile: Effects on Chilean Hake (*Merluccius gayi*)." *CalCOFI Reports* 49:157–66.

Alheit, Jürgen, and Miguel Niquen. 2004. "Regime Shifts in the Humboldt Current Ecosystem." *Progress in Oceanography* 60:201–22.

Aranciba, H., and S. Neira. 2008. "Overview of the Chilean Hake (*Merluccius gayi*) Stock, a Biomass Forecast, and the Jumbo Squid (*Dosidicus gigas*) Predator-Prey Relationship off Central Chile." *CalCOFI Reports* 49:104–15.

Ballón, Michael, et al. 2008. "The Impact of Overfishing and El Niño on the Condition Factor and Reproductive Success of Peruvian Hake, *Merluccius gayi peruanus*." *Progress in Oceanography* 79:300–307.

Brownlee, D. E., et al. 1997. "The Elemental Composition of Stony Cosmic Spherules." *Meteoritics and Planetary Science* 32:157–75.

Bustos, Claudia A., et al. 2007. "Spawning of the Southern Hake *Merluccius australis* in Chilean Fjords." *Fisheries Research* 83:23–32.

Escribano, Rubén, et al. 2004. "Biological and Chemical Consequences of the 1997–1998 El Niño in the Chilean Coastal Upwelling System: A Synthesis." *Deep-Sea Research II* 51:2389–411.

Gollner, Sabine, et al. 2017. "Resilience of Benthic Deep-Sea Fauna to Mining Activities." *Marine Environmental Research* 129:76–101.

Gutiérrez, Dimitri, et al. 2016. "Productivity and Sustainable Management of the Humboldt Current Large Marine Ecosystem under Climate Change." *Environmental Development* 17:126–44.

Hein, James R., et al. 2013. "Deep-Ocean Mineral Deposits as a Source of Critical Metals for High- and Green-Technology Applications: Comparison with Land-based Resources." *Ore Geology Reviews* 31:1–14.

Jurado-Molina, Jesús, et al. 2006. "Incorporating Cannibalism into an Age-Structured Model for the Chilean Hake." *Fisheries Research* 82:30–40.

Katija, Kakani, et al. 2017. "New Technology Reveals the Role of Giant Larvaceans in Oceanic Carbon Cycling." *Scientific Advances* 3:e1602374.

Li, Guancheng, et al. 2020. "Increasing Ocean Stratification over the Past Half-Century." *Nature Climate Change* 10:1116–23.

Miller, Kathryn A., et al. 2018. "An Overview of Seabed Mining Including the Current State of Development, Environmental Impacts, and Knowledge Gaps." *Frontiers in Marine Science* 4:418.

Montecino, Vivian, and Carina B. Lange. 2009. "The Humboldt Current System: Ecosystem Components and Processes, Fisheries, and Sediment Studies." *Progress in Oceanography* 83:65–79.

Oyarzún, Damián and Chris M. Brierly. 2019. "The Future of Coastal Upwelling from Model Projections." *Climate Dynamics* 52:599–615.

Payá, Ignacio, and Nelson M. Ehrhardt. 2005. "Comparative Sustainability Mechanisms of Two Hake (*Merluccius gayi* and *Merluccius australis*) Populations Subjected to Exploitation in Chile." *Bulletin of Marine Science* 76.2:261–86.

San Martín, Marcelo A., et al. 2013. "Relationship between Chilean Hake (*Merluccius gayi gayi*) Abundance and Environmental Conditions in the Central-Southern Zone of Chile." *Fisheries Research* 143:89–97.

Sweetman, Andrew K., et al. 2024. "Evidence of Dark Oxygen Production at the Abyssal Seafloor." *Nature Geoscience* 17:737–39.

Thiel, Martin, et al. 2007. "The Humboldt Current System of Northern and Central Chile: Oceanographic Processes, Ecological Interactions, and Socioeconomic Feedback." *Oceanography and Marine Biology: An Annual Review* 45:195–344.

Uścinowicz, Grzegorz. 2012. "Spherical, Magnetic Grains of Extraterrestrial Origin, Isolated from Pacific Sediments." *Oceanological and Hydrobiological Studies* 41.3:48–53.

Vanreusel, Ann, et al. 2016. "Threatened by Mining, Polymetallic Nodules Are Required to Preserve Abyssal Epifauna." *Scientific Reports* 6:26808.

Wedding, L. M., et al. 2015. "Managing Mining of the Deep Seabed." *Science* 349:144–45.

CHAPTER 13. DREAM OF THE GREEN TURTLES

Åkesson, S, et al. 2001. "Oceanic Long-Distance Navigation: Do Experienced Migrants Use the Earth's Magnetic Field?" *Journal of Navigation* 3:419–27.

Allen, M. S. 2007. "Three Millennia of Human and Sea Turtle Interactions in Remote Oceania." *Coral Reefs* 26:959–70.

Balazs, George H., and Milani Chaloupka. 2004. "Thirty-Year Recovery Trend in the Once Depleted Hawaiian Green Turtle Stock." *Biological Conservation* 117:491–98.

Broderick, Annette C., et al. 2006. "Are Green Turtles Globally Endangered?" *Global Ecology and Biogeography* 15:21–26.

Carr, Archie, and Jeanne A. Mortimer. 1987. "Reproduction and Migrations of the Ascension Island Green Turtle (*Chelonia mydas*)." *Copeia* 1:103–13.

Crawford, Sharika D. 2020. *The Last Turtlemen of the Caribbean*. Chapel Hill: University of North Carolina Press.

Darwin, Charles. 1873. "Perception in the Lower Animals." *Nature* 7 (March 13): 360.

Endres, Courtney S., et al. 2016. "Multi-Modal Homing in Sea Turtles: Modeling Dual Use of Geomagnetic and Chemical Cues in Island-Finding." *Frontiers in Behavioral Neuroscience* 10:19.

Fuentes, M.M.P.B., et al. 2010. "Past, Current, and Future Thermal Profiles of Green Turtle Nesting Grounds: Implications from Climate Change." *Journal of Experimental Marine Biology and Ecology* 383:56–64.

Glen, F., et al. 2006. "Thermal Control of Hatchling Emergence Patterns in Marine Turtles." *Journal of Experimental Marine Biology and Ecology* 334:31–42.

Godley, Brendan J., et al. 2001. "Nesting of Green Turtles (*Chelonia mydas*) at Ascension Island, South Atlantic." *Biological Conservation* 97:151–58.

———. 2002. "Temperature-Dependent Sex Determination of Ascension Island Green Turtles." *Marine Ecology Progress Series* 226:115–24.

Hamann, Mark, et al. 2013. "Climate Change and Marine Turtles," in *The Biology of Sea Turtles*, ed. Jeanette Wyneken, Kenneth J. Lohmann, and John A. Musick. London: CRC Press, 3:353–78.

Hays, G. C., et al. 2002. "Biphasal Long-Distance Migration in Green Turtles." *Animal Behaviour* 64:895–98.

———. 2003a. "Climate Change and Sea Turtles: A 150-Year Reconstruction of Incubation Temperatures at a Major Marine Turtle Rookery." *Global Change Biology* 9:642–46.

———. 2003b. "Island-Finding Ability of Marine Turtles." *Proceedings of the Royal Society of London B* 270:S5–S7.

Huxley, Roger. 1999. "Historical Overview of Marine Turtle Exploitation, Ascension Island, South Atlantic." *Marine Turtle Newsletter* 84:7–9.

Jones, K. 2016. "A Review of Fibropapillomatosis in Green Turtles (*Chelonia mydas*)." *Veterinary Journal* 212:48–57.

Lohmann, Kenneth J., et al. 2013. "Natal Homing and Imprinting in Sea Turtles," in *The Biology of Sea Turtles*, ed. Jeanette Wyneken, Kenneth J. Lohmann, and John A. Musick. London: CRC Press, 3:59–77.

Lorne, Jacquelyn Kay, and Michael Salmon. 2007. "Effects of Exposure to Artificial Lighting on Orientation of Hatchling Sea Turtles on the Beach and in the Ocean." *Endangered Species Research* 3:23–30.

McClenachan, Loren, et al. 2006. "Conservation Implications of Historic Sea Turtle Nesting Beach Loss." *Frontiers in Ecology and the Environment* 4.6:290–96.

Parsons, James J. 1962. *The Green Turtle and Man*. Gainesville: University of Florida Press.

Petry, Maria V., et al. 2021. "Plastic Ingestion by Juvenile Green Turtles (*Chelonia mydas*) off the Coast of Southern Brazil." *Marine Pollution Bulletin* 167:112337.

Pritchard, Adam M., et al. 2022. "Green Turtle Population Recovery at Aldabra Atoll Continues after 50 Years of Protection." *Endangered Species Research* 47:205–15.

Rizzi, Milena, et al. 2019. "Ingestion of Plastic Marine Litter by Sea Turtles in Southern Brazil: Abundance, Characteristics and Potential Selectivity." *Marine Pollution Bulletin* 140:536–48.

Schuyler, Qamar, et al. 2014. "Global Analysis of Anthropogenic Debris Ingestion by Sea Turtles." *Conservation Biology* 28.1:129–39.

Seminoff, Jeffrey A., et al. 2015. "Status Review of the Green Turtle (*Chelonia mydas*) under the Endangered Species Act." *National Oceanic and Atmospheric Administration* (NOAA).

Van Houtan, Kyle S., et al. 2010. "Land Use, Macroalgae, and a Tumor-Forming Disease in Marine Turtles." *PLOS ONE* 5.9:e12900.

Waycott, Michelle, et al. 2009. "Accelerating Loss of Seagrasses across the Globe Threatens Coastal Ecosystems." *PNAS* 106.30:12377–81.

Weber, Sam B., et al. 2012. "Fine-Scale Thermal Adaptation in a Green Turtle Nesting Population." *Proceedings of the Royal Society B* 279:1077–84.

———. 2014. "Recovery of the South Atlantic's Largest Green Turtle Nesting Population." *Biodiversity and Conservation* 23:3005–18.

EPILOGUE

Alroy, John. 2008. "Dynamics of Origination and Extinction in the Marine Fossil Record." *PNAS* 105:11536–42.

———. 2010. "The Shifting Balance of Diversity among Major Marine Animal Groups." *Science* 329:1191–94.

Danovaro, Roberto, et al. 2014. "Challenging the Paradigms of Deep-Sea Ecology." *Trends in Ecology and Evolution* 29.8:465–75.

Johnson, Kenneth G., et al. 2011. "Climate Change and Biosphere Response: Unlocking the Collections Vault." *BioScience* 61.2:147–53.

Jones, Daniel O. B., et al. 2014. "Global Reductions in Seafloor Biomass in Response to Climate Change." *Global Change Biology* 20:1861–72.

Lister, Adrian M., and Climate Change Research Group. 2011. "Natural History Collections as Sources of Long-term Datasets." *Trends in Ecology and Evolution* 26.4:153–54.

Mayhew, Peter J., et al. 2008. "A Long-term Association between Global Temperature and Biodiversity, Origination and Extinction in the Fossil Record." *Proceedings of the Royal Society B* 275:47–53.

McCauley, Douglas J., et al. 2015. "Marine Defaunation: Animal Loss in the Global Ocean." *Science* 347:1255641.

Molinos, Jorge Garcia, et al. 2016. "Climate Velocity and the Future Global Redistribution of Marine Biodiversity." *Nature Climate Change* 6:83–88.

Pinsky, Marlin, et al. 2013. "Marine Taxa Track Local Climate Velocities." *Science* 341:1239–42.

———. 2020. "Climate-Driven Shifts in Marine Species Ranges: Scaling from Organisms to Communities." *Annual Review of Marine Science* 12:153–79.

Poloczanska, Elvira S., et al. 2013. "Global Imprint of Climate Change on Marine Life." *Nature Climate Change* 3:919–25.

———. 2016. "Responses of Marine Organisms to Climate Change across Oceans." *Frontiers in Marine Science* 3:62.

Rillo, Marina C., et al. 2019. "Surface Sediment Samples from Early Age of Seafloor Exploration Can Provide a Late 19th Century Baseline of the Marine Environment." *Frontiers in Marine Science* 5. https://doi.org/10.3389/fmars.2018.00517.

Roemmich, Dean, et al. 2012. "135 Years of Global Ocean Warming between the *Challenger* Expedition and the Argo Programme." *Nature Climate Change* 2:425–28.

INDEX

Page numbers in *italics* refer to figures